THE FINAL EPIDEMIC
Physicians and Scientists on Nuclear War

edited by Ruth Adams and Susan Cullen

Published by the Educational Foundation for Nuclear Science
Chicago, Illinois 60637

Educational Foundation for Nuclear Science, Inc.
1020 East 58th Street,
Chicago, Illinois 60637
(312-363-5225)

This book was set in Times Roman by R&S Media Services and printed and bound by George Banta Company, Inc. in the United States of America.

Library of Congress Catalog Card No.: 81-69920
ISBN 0-941682 00-5

Distributed by:

The University of Chicago Press
Marketing Department
5801 S. Ellis Avenue
Chicago, Illinois 60637
(312-962-3510)

UCP Order No.: 03874-2

Drawings by Robert Jay Lifton are from *Birds* (Random House, N.Y., 1968) and *PsychoBirds* (Countryman Press, Taftsville, Vt., 1978).

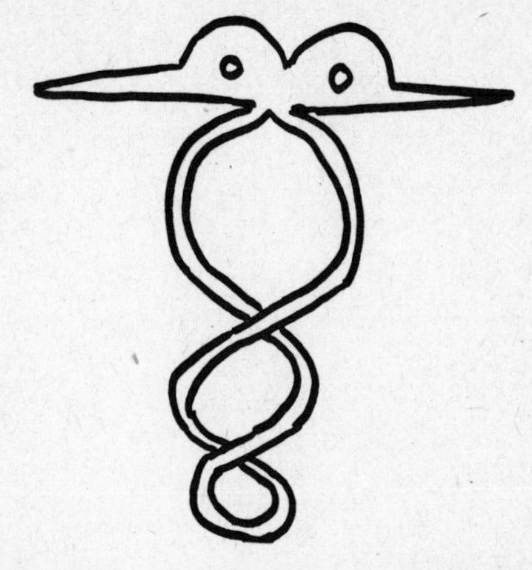

"An inescapable lesson of contemporary medicine is that when treatment of a given disease is ineffective or where costs are insupportable, attention must be given to prevention."—Howard Hiatt, M.D.

Contents

THE FINAL EPIDEMIC was a project undertaken in cooperation with the Physicians for Social Responsibility and the Council for a Livable World Education Fund. Publication of the book was supported by a grant from the J. Roderick MacArthur Foundation which we gratefully acknowledge.

Preface

GEORGE B. KISTIAKOWSKY

"Over all these years the competition in the development of nuclear weaponry has proceeded steadily, relentlessly, without the faintest regard for all the warning voices. We have gone on piling weapon upon weapon, missile upon missile, new levels of destructiveness upon old ones. We have done all this helplessly, almost involuntarily, like the victims of some sort of hypnotism, like men in a dream, like lemmings heading for the sea, like the children of Hamlin marching blithely along behind their Pied Piper."—Ambassador George Kennan, May 1981

Leo Szilard, the farsighted physicist, who in 1940 induced Albert Einstein to warn President Roosevelt that nuclear fission, newly discovered in Nazi Germany, made possible the making of bombs of hitherto unimaginably destructive power, devoted his energy and enterprise after World War II to efforts against the nuclear arms race engaged in by the United States and the Soviet Union. Ill, and despairing of his efforts to reverse the arms build-up policies of the two superpowers, he founded in 1962 the Council for a Livable World, dedicating it to combatting the menace of a nuclear war.

Since then several international agreements and treaties concerned with nuclear arms, bilateral and multilateral, have come into force; but their effect has not been to control the arms race. At most they have created a framework within which this race proceeds with little hindrance. Some optimists have argued that progress in arms control is being made and that an even greater barrier to the use of nuclear weapons exists because of the passage of time, now 36 years since the first and last use of them in warfare. But is this trend real? Does the passage of time make the world safer from nuclear holocaust? The evidence, I fear, points the other way.

Until the last decade the emphasis was on creating a deterrent against attack. In the main the weaponry was supposed to be a defensive force threatening assured destruction of unacceptable scope, politically and economically, to the society of the aggressor. In the 1970s, however, came the deployment of MIRVs (multiple independently targetable reentry vehicles) for long-range missiles and the growth of Soviet nuclear forces, which began to approach and in some aspects to surpass those of the United States. These developments led to doubts in the frequently paranoid minds of militarists about the political effectiveness of nuclear deterrence. Proposals then surfaced for being prepared to fight protracted and "controlled" nuclear wars by attacking assorted military targets, not the industrial population centers. Proponents of the use of military force in the conduct of foreign policy engaged in intensive propaganda—nuclear war is not the ultimate disaster; it is winnable if you are strong; your chance for survival is really not so bad.

The result has been a great renaissance of militarism in the American populace. Coupled with exaggerated fears of the Soviet Union, it has provided a solid foundation for an aggressive and even reckless foreign policy. Assertions about winnable and controlled nuclear wars, however, are disputed by experienced former Pentagon officials and by quite a few senior military figures here and abroad, who are all convinced that a nuclear war between the superpowers cannot be controlled and will inevitably escalate into a holocaust destroying the civilizations of the United States and the Soviet Union as well as doing great damage to the rest of the world.

A succession of political and military actions by both superpowers has caused their relations to worsen greatly since the days of burgeoning detente some ten years ago. The major expansion of our military budgets by President Carter, now escalated to gigantic proportions by President Reagan, and the barely veiled threats of invasion of Poland by the Soviet Union are bringing back the Cold War of the 1950s. Not surprisingly, all of this stimulated the militarization of the rest of the world. Total military annual expenditures have risen well past $500 billion, which is more than is spent on education or public health. Over 100 million individuals are now paid directly or indirectly by defense ministries: 25 million are in the world's regular military forces; others are members of paramilitary or reservist forces, civilians directly employed by the military or those working on weapons research production or related activities.

The huge trade in modern military hardware is a severe drain on the resources of the Third World, in addition to that caused by the massive rise in the price of energy. Unfortunately one can anticipate only added tensions, some caused by future shortages of non-renewable natural resources, others by the failure of agricultural production to keep pace with the growing world population, hundreds of millions of whom already exist on the very edge of starvation.

The end of World War II did not bring peace to the world. A study published by the Brookings Institution determined that in the period 1946 to 1975 there were 215 armed conflicts: foreign wars, civil wars, insurrections and organized guerrilla fighting. More have taken place since 1975, and as of now at least 8 such conflicts are underway, including the warfare in Afghanistan between the local guerrillas and the joint forces of the Soviet Union and its puppet government, as well as the war between Iraq and Iran. In most of these conflicts the superpowers are involved by proxy, as extensive use is made of their modern high firepower hardware, provided either free or on very easy terms. Repeatedly the superpowers felt threatened, as when, for instance, the American Strategic Air Command with its nuclear warheads was put on alert 33 times in the period ending in 1975.

Fortunately, no nuclear weapons have been used in warfare since 1945, but the efforts to prevent their spread to other nations have not been successful and the Non-Proliferation Treaty, in force since 1970, is clearly failing. The responsibility is shared by the superpowers which set an example by failing to carry out their treaty commitment to make determined efforts at nuclear disarmament, and instead greatly expanded their nuclear arms. There are now six states which admit having exploded a nuclear device, although India claims that its was a peaceful explosion. Israel is authoritatively said to be in possession of nearly ready-to-use warheads, and its raid on the nuclear reactor in Iraq has set a dangerous precedent for other nations. Iraq, Pakistan, Libya, South Africa, Argentina, Brazil and perhaps other states are in various stages of acquiring nuclear warheads. An authoritative Defense Department/CIA group predicted in 1979 that in the 1990s there will be at least two dozen nuclear-armed states in the world.

Given the frequency of local wars and the quality of the political leadership in so many lands, the conclusion is virtually inescapable that somewhere, sometime in the foreseeable future one or more atom bombs will be exploded in warfare. That this local disaster will engulf some of neighboring states is far from improbable. In fact, considering the present trends, there is no assurance that the United States and the Soviet Union will not enter the fray and so set the stage for a holocaust, like lemmings marching into the sea, in the words of George Kennan.

The augury for the future is not a happy one. Painfully, one must admit nearly total failure of the continuing efforts of peace-oriented organizations to ward off the threat of nuclear war. Among them is the Council for a Livable World, which following the guidance of Szilard, concentrated its efforts on raising financial assistance to senatorial candidates dedicated to nuclear arms control, in the conviction that the make-up of the U.S. Senate is critical to the preservation of peace. The Council has also arranged regular and quite well attended off-the-record presentations to members of the Senate to clarify the technical and other aspects of specific proposals for arms control and to assess the impact of new weapons systems proposed by the military.

The political events of the last few years have demonstrated, however, how effective can be the electoral campaigns of right-wing and related single-issue organizations, which have now come to influence strongly the makeup and therefore the decisions of the Senate. To oppose effectively and reverse these trends calls for an equally aggressive, direct campaign leading to a broad-based public movement for peace. With this in mind the Council established in 1979 an affiliated Council for a Livable World Education Fund. One of the major activities of this group became its participation in the symposia on "The medical consequences of nuclear weapons and nuclear war," undertaken by Physicians for Social Responsibility. While the symposia attract mainly a medical audience, the Council and the Physicians for Social Responsibility believe that our messages will be carried by physicians to broader strata of the public and thus will be a significant contribution to the mass movement needed to prevent the start of the last epidemic, nuclear war between the superpowers.

Introduction

HELEN CALDICOTT

The world is moving rapidly toward the final medical epidemic, thermonuclear war. This planet can be compared to a terminally ill patient infected with lethal "macrobes" which are metastasizing rapidly. The terminal event will be essentially medical in nature, but there will be few physicians remaining to treat the survivors.

In the past, incurable epidemics of bacterial or viral disease have been controlled or eliminated by preventive medicine. The lessons of preventive medicine apply equally to the most serious event threatening to befall the human race—nuclear war. It can be averted by a concerted and urgent campaign conducted by physicians throughout the world, united by an allegiance to the ancient Hippocratic oath.

Such a campaign has already begun. This book is a unique collection of papers presented at a series of symposia on the medical consequences of nuclear weapons and nuclear war, organized by Physicians for Social Responsibility and the Council for a Livable World Education Fund. The authors are persons of international stature in medicine, science and politics, drawn together by their concern for survival.

Beginning in 1979, these symposia have been conducted in Boston, New York, New Haven, San Francisco, Seattle, Chicago, Albuquerque and Los Angeles and more are planned. They are sponsored by prestigious medical schools and, frequently, state medical societies. Physicians attending them have been accorded Continuing Medical Education credits. Extensive media coverage, both local and national, has evoked a positive response from the public, and from many government officials who are thinking about this subject for the first time since the Cuban Missile Crisis and the period of atmospheric nuclear weapons testing and public debate about fallout shelters.

Physicians for Social Responsibility was initiated in 1961 by a small group of doctors in the Boston area, concerned about the effects of atmospheric nuclear tests. They wrote a two-part article, "Human and Ecologic Effects in Massachusetts of an Assumed Thermonuclear Attack on the U.S." and "The Physician's Role in the Postattack Period" for the *New England Journal of Medicine,* 266, 22 (1962), pp. 1127-45, dealing with the medical consequences of nuclear war. These essays, the first of their kind, prompted requests for many reprints, including hundreds ordered by the Department of Defense. The data they set forth, even more topical in 1981, played an important part in the debate that led to the Limited Test Ban Treaty in 1963.

In 1979, Physicians for Social Responsibility was reinvigorated by a small group of physicians. We now have more than 5,000 members with almost 50 chapters, operating or in formation. These physicians are taking the responsibility to explain the medical implications of the nuclear age. They are involved in public speaking, media appearances, distribution of literature, congressional debates and hearings, and education of citizens about weapons facilities and nuclear installations in their communities.

We believe that the key to survival is in our hands. Historically, medical knowledge has been instrumental in the formulation of local, national and international government policy. Scientific descriptions of bacterial and viral diseases initiated the era of preventive medicine, with the organization of sanitation systems, uncontaminated water and food, personal hygiene and huge immunization programs.

Education was adequate for the formulation of these policies, but now education alone is not enough.

Many people seek to escape the realities of the dangers that nuclear weapons pose by ignoring them. Robert Lifton, a psychiatrist who has studied the psychological effects of the nuclear age, has called the process "psychic numbing." Overcoming this process will not be easy, but it is essential if people are to come to grips with the nuclear threat.

Society looks to physicians for guidance on issues of life and death. When their warnings are heeded and patients begin to look realistically at the threats they face, they may find the prospect terrifying, but this experience is essential to the emotional adaptation necessary for effective political action. This book will serve as a powerful tool for people striving to achieve this goal.

Time is short. The prognosis is guarded. But the analogy of the terminally ill patient in the intensive care unit who occasionally survives because of dedicated medical, scientific and nursing efforts seems appropriate.

Nothing less will save this Earth. Physicians daily practice the art of medicine. Let us join together, doctors and patients, to preserve what may be the only life in the universe. We are at the crossroads of time. Only emotional maturity, evoked by extreme danger combined with personal responsibility and total commitment, will save our planet for our descendants.

I SYMPTOMS

"Those who say that we should accept the risk of nuclear conflict to save our system are saying, in the strongest possible terms, that we should accept its certain destruction."—John Kenneth Galbraith

What are your scholarly conclusions concerning the biomedical and psychosocial effects of nuclear war?
Ugh.

1 In a dark time . . .

ROBERT JAY LIFTON

Physicians today must extend their healing imagination to unprecedented threats to human life, threats to the human species. This is not easy. We are much more comfortable in a one-to-one situation of individual doctor and patient.

I myself have strayed a great distance from the medical heartland—not only as a psychiatrist, but as a psychiatrist concerned more with historical than clinical matters. Yet wherever a physician strays, he takes some of that healing responsibility with him.

At this moment in history, physicians can't afford to suppress that responsibility. If we do not contribute our knowledge to the prevention of nuclear war, then where will our offices be, or our clinics or hospitals? Where will we practice our one-to-one healing skills? Dick Gregory was once asked, at an anti-nuclear rally, "What are you doing here? Why aren't you out demonstrating for black rights?" He replied, "Look, if I demonstrate for black rights in the morning and they drop the bomb at noon, there won't be any rights to demonstrate for in the evening." It's the kind of situation we are all in, whatever our professions.

Recent work I've done with former Nazi doctors has, I confess, sensitized me to questions about healing and killing. In their case the perversion was so great as to create a transformation from healer to killer. For most of the rest of us, however, the danger is more that of standing aside, resting on our virtue as healers, even as the world moves closer to destroying itself.

The Hiroshima experience cannot really represent what would happen to people if our contemporary nuclear weapons were used. Today's weapons have hundreds or thousands of times greater destructive power, and the potential is simply not comparable.

Still, even in that tiny Hiroshima bomb experience, there are indications of the psychology of the nuclear age, of what I sometimes call Nuclear Man. These indications have a lot to do with questions of boundaries and the new boundarylessness of human destruction. That is why I called my book about Hiroshima *Death in Life*. The death that survivors carried on in their continuing lives was not the ordinary structured death of individuals; it was grotesque, absurd, collective, unacceptable, unabsorbable death.

That's what the politicians don't talk about. It isn't so much what Krushchev said—that the survivors will envy the dead. It is rather that the survivors themselves will be as if dead.

In a split second of being exposed to that first atomic bomb dropped on a human population, everybody in Hiroshima experienced a permanent encounter with death. We can think of that encounter as taking place in four stages and becoming endless.

• The first stage is obviously the moment of the bomb's fall, the overwhelming emergence of death, of the dying and near dead around one. Even the estimates of the number killed are indicative, because nobody really knows how many people were killed in Hiroshima: anywhere from around 60,000 to 300,000. The city of Hiroshima estimates 200,000. It depends upon how you count, which groups you count, whether you count deaths over time. And it depends on emotional influences on the counters. It is of some significance that U.S. estimates have tended to be lower.

The real point is that even that small event doesn't lend itself to calculations of the degree of destruction. Certainly, all of Hiroshima immediately became involved in that atomic holocaust. No one relatively near the center of the city would escape from the sense of ubiquitous death. So I characterize as the most significant psychological feature of this first stage the sudden and absolute shift from normal existence to an overwhelming encounter with death.

The main function of what I call psychic numbing is diminished capacity or inclination to feel, to take into account the experience of what happens at the receiving end of the weapon. When people described this to me, some said, "My first feeling was: I think I will die, this is the end for me." But it was more than the idea of individual death.

In Hiroshima those early images included something like the idea that the whole world was dying. As one professor put it: "My body seemed all black, everything seemed dark, dark all over; then I thought, the world is ending." A Protestant minister, responding to the scenes of mutilation and destruction as he walked through the city, said: "The feeling I had was that everyone was dead. The whole city was destroyed. I thought that all my family must be dead. It doesn't matter if I die. I thought this was the end of Hiroshima, of Japan, of humankind." And a woman writer: "I thought it must have been something which had nothing to do with the war; the collapse of the earth which is said to take place at the end of the world which I had read about as a child."

The related feeling that survivors conveyed to me was confusion over who was alive and who was dead. Many said, "I did not think I was really alive," or they spoke of human ghosts, people who walked in the realm of dreams. The breakdown of the dividing line between life and death was manifest.

- The second stage of this permanent encounter with death was what I came to call invisible contamination—the acute impact of irradiation. Unprecedented and totally unknown, it was grotesque and mystifying to the people of Hiroshima to see in themselves and others, whether within minutes or hours, or over days, weeks, or months, the symptoms of acute irradiation. These can be gastrointestinal, with anorexia, severe bloody diarrhea, weakness, high fever. There can also be bleeding into the skin and from all orifices. Later manifestations were disfiguring symptoms, including the loss of all body hair. And of course, very low white counts were observed when blood tests could eventually be made.

This gave the people in Hiroshima the image of a weapon that not only kills and destroys on a colossal scale, but also leaves behind, in the bodies of those exposed to it, deadly influences which may at any time strike down their victims. And that second stage is epitomized in a series of rumors that swept Hiroshima soon after the bomb fell.

One was perhaps predictable: that everyone who had been exposed to the bomb would be dead within a certain time, say two or three years; or that the city would be uninhabitable for as long as 75 years. But there was a second rumor, even more frequently reported to me, and perhaps with even greater emotion: the belief that trees, grass and flowers would never again grow in Hiroshima. Here the underlying meaning was that life was being extinguished at its source, an ultimate form of desolation that seemed to encompass but go beyond human death.

• The third encounter with death had to do with delayed radiation effects, not months but years after the atomic bomb itself. In the late 1940s and early 1950s, people began to talk about what they called A-bomb disease. That condition has no scientific standing, but it was terrifying psychologically. The model for what people called A-bomb disease was leukemia, because it was discovered within three years or so that the rate of leukemias had increased among those significantly exposed to irradiation in Hiroshima. Many leukemia symptoms are very similar to those in acute irradiation, so people had the sense of the bomb's effects continuing indefinitely, and of always being lethal.

Then, sometimes decades later, there began to emerge increased incidence of various forms of cancer. Thyroid cancer showed earlier increases, but afterwards there was more cancer of the breast, lung, ovary, and other areas. These cancers are not very rare; but now, even if the incidence is only slightly increased, it again creates the sense of endless lethal influence, the feeling the bomb can do anything and anything it does is likely to be fatal.

There also is a vast medical or psychosomatic gray area. Many of the medical issues relating to Hiroshima are still uncertain, and we do not have a complete grasp of them. But in the areas of blood disease, endocrine and skin disorders, damage to the central nervous system, premature aging, impairments in early development and a vague borderline condition of general weakness and debilitation, there are positive findings by some observers. True, there is often a lack of systematic confirmation in controlled populations: nobody is quite sure. But there is a psychological bind for the survivors because any kind of condition, from fulminating leukemia to a common cold, creates the immediate association with A-bomb disease, with bomb fatality.

Another issue is that of genetic effects, again a controversial one in Hiroshima. The lay population of Hiroshima confuses genetic effects with conditions that have nothing scientifically to do with them, specifically, the damage to fetuses or embryos in utero. This damage, of course, results from the destructive effects of irradiation on sensitive, rapidly growing tissue. There were, of course, striking microcephalic and other forms of mental retardation directly attributed to radiation effects from the bomb, and these were immediately associated with genetic effects.

There is still much concern about second generation survivors in terms of potential genetic effects. Another theory is that the second generation may not be affected but the third generation may be. Can anybody know? We do know that genetic effects can occur and that nobody can guarantee that survivors of Hiroshima or their descendants won't be affected. This, too, gives people a sense of an endless chain of lethal possibilities: if ill effects don't appear in one year or in one generation, they may well appear in the next.

• The fourth—or really the continuing—endless stage is that of, simply, the identity of atomic bomb survivors. I began to call it an identity of the dead or of the people who feel dead. Many survivors did recover to a considerable degree. Many worked and married, although they also met discrimination about jobs and marriage under the "rational assumption" that, being weakened through irradiation exposure, they would make less reliable workers, or would be defective husbands or wives genetically. But beneath those reasons was the sense that these survivors carried a special death taint that was threatening. People wanted to move away from them, to stay away from them.

It is as if the victim had internalized or absorbed the horror and evil of that which had victimized him—a very cruel psychological fate. There has not been a capacity among the people of Hiroshima to deal adequately with the experience, with their losses; in short, to commemorate the experience in a satisfying way.

They have tried many things. There have been survivor missions of great importance in which people have taken the responsibility of telling the world about nuclear war in order to give some meaning to what they have been through. These attempts have sometimes ended in conflict and political controversy.

There is no model. There is no proper way to behave for an individual or city that has felt the effects of an atomic bomb. This has to do with the destruction of the social fabric; even though the city of Hiroshima has been rebuilt, the social fabric of that community of survivors has not. A single small weapon has created a totality of destruction. There is unending lethal influence, a sense of being a victim of a force that threatens the species, that reverberates psychologically on those same people. That is quite a lot for this tiny first weapon of the nuclear age.

The point is sometimes raised that the people of Hiroshima were worse off because they had no prior knowledge of the bomb's effects. If it happens now, people will know it is irradiation. Will that help? It is still invisible contamination, and knowing its deadly nature may be just as painful as the terrible mystery that afflicted Hiroshima. Should the bomb actually fall, being informed does not seem a great advantage.

What was crucial was the existence of an outside world to help. People could come from outside of Hiroshima. Some doctors

came from nearby cities. And when the occupation forces appeared, they were helpful too. The help was sporadic and inadequate at the beginning; people died from lack of medical care over the first years. But still, the outside world was there. With today's nuclear weapons, can there be any assurance of this source of help? I don't think so.

Edward Teller writes in his extraordinarily misleading book, *The Legacy of Hiroshima*, that "rational behavior" consists of having the "courage" to use nuclear weapons when tactically indicated, and being "prepared to survive an all-out nuclear attack." My response is that this rational behavior is the logic of madness, and deadly madness at that.

Similarly, Herman Kahn, in his earlier study on thermonuclear war, tells us that a reasonable, or non-hypochondriacal individual who survives a future nuclear war "should be willing to accept almost with equanimity, somewhat larger risks than those to which we subject our industrial workers in peace time." He also talks about preventing undue fear of irradiation right after the nuclear attack by having available individual meters, a variety of geiger counters, to measure everybody's irradiation levels. He gives the example of a man who is ill from something, presumably other than irradiation: "You look at his meter and say, you've only received ten rems, why are you vomiting? Pull yourself together and get to work."

Here I must tell Herman Kahn that nuclear weapons make hypochondriacs of us all, if we survive. His wishful scenario of recovery has little to do with the way people behave, and his psychological assumptions are as faulty as his moral ones.

Some things are happening that raise real hope for a change in consciousness about nuclear weapons.

In the mid-1940s there came into the world a new image, that of exterminating ourselves as a species with our own technology. This image had been put forward by earlier visionaries like H.G. Wells. And indeed Leo Szilard realized that a nuclear weapon was possible after reading Wells' fiction during the mid-1930s. Of course, Szilard knew a few things in addition to what Wells had written. In 1939 he wrote to the 10 or 15 physicists in the world who were working in atomic physics and asked for a moratorium on publication of papers that might be potentially useful in the making of a nuclear—or atomic—weapon.

Although the idea of apocalypse has been with us throughout the ages, it has been within a religious context—the idea that God will punish and even eliminate man for his sins. Now it is our own technology, and we are doing it ourselves. Nor is it only the nuclear threat. There are chemical warfare and germ warfare, destruction of the environment, the air we breathe or the ozone layer, the depletion of the world's resources, whether of energy or food.

All of these have given us an image of potential extinction and tend to merge psychically in our minds. What we need is a psychology that moves away from instinct and defense toward symbolization of life and symbolization of death.

We all as human beings have not just immediate psychological responses, but also what Paul Tillich called ultimate concern—or what I call the symbolization of immortality. This is not denial of death; it is part of the fact that we know that we die. We also know that we live as cultural animals in some sort of continuum with those who have gone before and those who will go after our finite life span. So we seek a sense of living on in our sons and daughters. That is the biological mode of symbolizing immortality.

There is also a creative mode of influence, of living on in what we convey to our students, our readers or just to our friends. This is the human influence that we leave behind us.

A third mode is a religious one—the concept of a spirituality in which death is in some degree conquered—which may or may not postulate something like life after death.

A further mode is that of living on in eternal, unending relationship to nature. And finally, there is the traditional mode of the mystics, the psychic experience so intense that time and death disappear.

These are all part of our normative and necessary functioning. They are left too much to the philosophers and theologians, but are very much a matter for psychology. If we consider the possibility of nuclear extermination or near-extermination, who can then believe in living on in our sons and daughters and their sons and daughters in the biological mode; or in living on through our works? What works will continue, what influences?

Even in a theological sense, it would seem that, as the people in Hiroshima repeatedly told me, there was no idea powerful enough to help them accept what had happened to them. It may be that religious ideas, to be effective, require a sense of being among the living. And because we are capable of destroying the environment with our technology, even nature is not assuredly there for us.

Many of the young and not so young, who made the effort to move into a communal life in nature, were not seeking only a solution for a post-industrial society. Rather, they were trying to recapture a human relationship to nature which they felt was being threatened.

All of these modes are not lost to us totally; they are in doubt. And it is perhaps that doubt about human continuity that has given such great emphasis in recent decades to the last mode I described: the experience of transcendence. Perhaps that has a lot to do with the seeking of highs through drugs or through meditation or, for that matter, the appearance of the extremist religious cults which offer many kinds of highs and give a kind of cosmology that in many cases takes into account the nuclear threat.

There is some recent new work that suggests a way of getting at some of these psychological reactions as they affect specific people. Michael Carey, a former assistant of mine, did a study in the mid-1970s of people of his generation, then in their late twenties and early thirties.

He started by interviewing them about those quaint air-raid drills of the 1950s, in which students were taught that there was a terribly dangerous weapon but that you could protect yourself from it if you put your head under the desk or put a piece of paper over your head. Perhaps this was part of what Edward Teller called preparation for all-out nuclear attack.

Carey then questioned them about general ideas such as death and dying, human continuity, and other concepts stemming from these. Very simply, his findings, confirmed by others studying related matters, were:

- first, there was an early experience of absolute terror, of dreams, fears, fantasies of total destruction involving family, parents, oneself.
- that was suppressed fairly soon, often in later childhood: a kind of period of suppression or latency.
- later on, often in late adolescence or in adult life, there would be some form of return of what had been, or the emergence from psychic numbing, with all kinds of dreams and fears.

Sometimes people divide themselves into those who think a great deal about these things and those who don't. But almost every person interviewed produced, after just one or two questions, a flow of associations. Nor did it make a difference whether the interviewee defined himself or herself as the kind of person who was concerned about these matters.

Certain themes emerged, one of them being the equation of death with collective annihilation. In early childhood, we all began to learn the terrible lesson that death is forever: when you die you don't lie down and then get up, as children like to play in games. But, for many in our culture, learning about death is simultaneous with the association of annihilation. That, I think, is an enormous deformation which we have hardly begun to think about. It may be the fundamental deformation of the nuclear age.

A second theme in Carey's work and other studies is that of pervasive doubts about the lasting nature of anything. It might be labeled the new ephemeralism: nothing can be trusted to last. I think this is related to attitudes toward work, ranging from extreme hedonism, the so-called hippy ethos, to a committed focus on a narrow definition of work.

A third theme is the perception of craziness. Of course, those six-year-old kids were much too intelligent to believe that putting a piece of paper over their heads would protect them from the bomb. This perception had much to do with a sense of absurdity in that generation, and it can be a healthy sense. The whole idea of combatting the bomb seemed crazy; the authorities seemed crazy; the world seemed crazy, as indeed in many ways it is.

There was even the theme of identification with the bomb. Some of the people in Carey's study wanted to see it go off, because they were curious about it, or because it gave them a sense of power to imagine the bomb going off. That is a dangerous one, unfortunately present in certain highly placed people in our society.

A final theme is that of a double life, which we all lead in this nuclear age. When we think about it, we know that in a moment, through miscalculations, error, the wrong kind of nationalism or ideological excess, a nuclear war may start; most of humankind may be exterminated. Yet we also go on living our ordinary lives, doing our work, going about our ordinary everyday matters as if there were no such threat.

In a social and perhaps a political sense, we are in a revolutionary situation without an appropriate response. It became evident in Carey's study that the double life is immobilizing. The capacity to do something, to release the energy needed to act on these issues, would be enormously beneficial—potentially to the world, but also to people caught up in this immobility.

Two other deformations deserve attention. One is the psychic numbing to which I have already referred. It has to do with diminished capacity or inclination to feel. It can take the form of a blocking of feelings or images, or both. People in Hiroshima could say to me: "You know, I could see what was going on around me, I could see people dead or dying, but suddenly I felt nothing." But this psychological blocking occurs in all of us, and certainly in those who make the weapons or project their possible use.

Worse than that is what I term the religion of nuclearism, in which we embrace the very agents of our potential extermination as a source of salvation. Herman Kahn and Edward Teller exemplify an ideology in which the weapons themselves become a solution to death anxiety, to restoring a lost sense of immortality.

Robert Oppenheimer thought that the dropping of the bomb would teach mankind a lesson, so that maybe there would never be wars again. We also tried to use the bomb diplomatically, to attain goals we could not otherwise achieve, to replace human efforts, psychological and human skills. Some of the reluctance to face the enormous dangers inherent in nuclear power stems from this deification of the nuclear weapon. The deity, if it can destroy, can also create a world of milk and honey.

There can also be a kind of nuclear backsliding. Oppenheimer, who was very committed to using the first bomb, finally came to a sense of revulsion—not with the beginning of the atomic bomb but with the large thermonuclear bomb. He then wrote brilliantly on its dangers, not just to us as a species but also to our minds. But when he underwent this reversal he was crucified.

There are devil's bargains in all cultures. An American example was closing our eyes to holocaust. That has changed, but we tend more to talk about holocausts of the past. We don't talk much about the danger of future holocausts. And when we do, there is an enormous effort to domesticate holocaust and the experience of survivors. We are familiar with the bland language of nuclear weapons: exchanges, stockpiles, scenarios.

Sometimes, however, events break through the devil's bargain. One was the Three Mile Island accident and the other, seemingly unrelated, the showing of the film, "Holocaust." Both evoked extraordinary responses, and my point is that the two events didn't create these anxieties and responses. They tapped pre-existing anxiety and a sense of terror around this imagery of extermination.

In such anxiety we can find at least the seeds of a more formed awareness. We need a certain amount of anxiety or tension to overcome the numbing. The kind of psychic numbing that makes us comfortable and that we have always resorted to functions to keep us from thinking about these things. But that very kind of numbing is enormously dangerous collectively to the survival of the species.

There is a real conflict. The ideal level of tension is hard to achieve, but we must constantly seek it. To act effectively, it isn't necessary first to realize all the terrible things that the nuclear weapons can do. Of course, we must eventually come to that realization. But the other way around also works: We need some kind of humane standard. We have to remember that it is wrong, it is evil to engage in genocide or to project genocidal "scenarios." And then, when we have achieved that simple kind of moral perspective, we can begin to take in some of the realities of what does happen at the other end of these weapons. We are numb to reality without that simple act of moral imagination.

A potential change of consciousness is held out before us. The responses to Three Mile Island in particular are strongly resisted; there is a psychological and economic tendency in which the more evidence there is of the extraordinary danger of nuclear power, the more effort there is to deny that evidence.

But in my view the increasing awareness of the danger of nuclear power has enormous bearing on the nuclear weapons issue—not only because of the technological relationships, but because nuclear energy gives us a more direct perception of threat. The "wisdom of the body" senses that this stuff can affect our concerns about nuclear weapons.

Perhaps it is too soon to speak of a shift in consciousness having occurred. Perhaps it is a turn toward awareness, just as we used to speak of a turn toward peace in connection with the Vietnam war. But that is a great deal. Certainly, we have to agree that however things go, we are all involved in the process.

There is a wonderful line from Theodore Roethke, the American poet, which has haunted me since I read it some time ago: "In a dark time, the eye begins to see."

2 Psychosocial trauma

JOHN E. MACK

How can we create a climate in which there can be citizen participation in the decisions that affect the nuclear arms race—the most critical issue of our time? And what has interfered with the involvement of myself and others?

I have been warned by non-medical colleagues that there is a danger for a psychiatrist in getting into this subject, that people will fear or assume that diagnoses or interpretations are going to be offered at a personal level. I am not going to examine individual aggression or personal motives, and surely not individual psychopathology. My ideas insofar as they apply to individuals do so only because individuals are caught up in systems of thought.

There are fundamental ways in which the nuclear arms race is psychological at its roots. This is true above all in the area of strategy. Military strategies and policies are based on what one perceives the intentions of an adversary to be, and upon what each believes will be the impact of a particular policy on the minds of those making decisions on the other side. There is a madness in the spiraling arms escalation. But where does it lie and where might we put our efforts in trying to get out of the box in which we are all trapped?

I participated in a Task Force of the American Psychiatric Association formed in 1977 at the urging of Perry Ottenberg, a psychiatrist who has been devoted to the study of social problems. The charge to the Task Force was to bring psychological understanding to bear on various aspects of the development of nuclear arms and nuclear energy, and the threat that they pose to human mental and emotional life.

My initial reaction to the invitation to take part in the Task Force was perhaps not atypical. I thought this was certainly an important problem, but why now? I found this response was also characteristic of my friends and psychiatric colleagues. But the events of the past few months have created a sense of urgency. At the same time the emphasis in the Task Force has shifted from concern with nuclear power production and its risks to the far more ominous danger of the arms race. I have also found that the resistances to becoming involved noted in myself are shared by the individuals and groups with whom I have discussed this problem. There is a gut response of "Oh my God, that's such an awful subject," and a tendency to want to "leave it to the experts" who understand the technological aspects of the problem.

One fact, however, emerges. Behind the technical language, the jargon, the acronyms, graphs, diagrams, statistics and strategic discussions, the concepts behind the arms race are not arcane or difficult. Some of the technology is, of course, complicated, but not the thinking related to its use.

The Task Force has been studying the psychological effects upon children and adolescents of living in a world where thermonuclear disaster is a constant threat. The subject has been little studied. What work has been done demonstrates that children are aware of the threat of nuclear war and live in fear of it. The data for this study have been gathered by William Beardslee and myself with technical help from a consultant, James Henning.

A questionnaire was given to 1,000 grammar and high school students between 1978 and 1980. More detailed responses were obtained from 100 students (10th to 12th grades) in the Boston area who were attending two schools, one a public school about 30 miles north of Boston, the other a private school 10 miles west of the city. The students were asked such questions as "What does the word nuclear bring to mind?" "How old were you when you were first aware of nuclear advances?" "Have you participated in any activity related to nuclear technology?" "Do you think you

could survive a nuclear attack?" "Have nuclear advances influenced your plans for marriage, having children or planning for the future?" and "Have nuclear advances affected your way of thinking about the future, your view of the world, and time?" We were aware that the use of the word "advances" is in one sense euphemistic, reflecting a kind of obliqueness of approach to a topic which includes the production of weapons that threaten human annihilation.

The students given the questionnaires probably represent a somewhat biased sample, as they are better informed about the dangers of nuclear war than we would expect the average adolescent school population to be (whatever may be meant by "average"). The comments are, nevertheless, quite disturbing and demonstrate that the imminent threat of nuclear annihilation has penetrated deeply into their consciousness. Some of their responses are as follows:

What does the word nuclear bring to mind?

"Danger, death, sadness, corruption, explosion, cancer, children, waste, bombs, pollution, terrible, terrible devaluing of human life."

"Nuclear means a source of energy which could provide the world with energy needed for future generations. It also means the destruction of marine life whose environment is ruined by nuclear waste. Also the destruction of human life when used in missiles."

The great majority did not believe that they, their city or the country could survive a nuclear attack.

The comments below are characteristic of the ones received in response to the question.

Have nuclear advances influenced your plans for marriage, having children or planning for the future?

"I don't choose to bring up children in a world of such horrors and dangers of deformation. The world might be gone in two seconds from now, but I still plan for the future, because I am going to live as long as I am going to live."

"Nuclear advances are not always on my mind. My philosophy is that life is full of dangers and troubles and worries—I can't spend my time on earth a psychologically sick person, afraid that at any moment I will die. I feel that I would refrain from having children, though, not because of thermonuclear threats—because I am not crazy about children."

"No, not really because if there is a nuclear war there is no sense in worrying about it because whatever happens will happen. The technology is there and it can destroy the world."

Have nuclear advances affected your way of thinking about the future, your view of the world, and time?

"I am constantly aware that at any second the world might blow up in my face. It makes living more interesting."

"I don't really worry about it, but it is terrifying to think that the world may not be here in a half hour. But I am still going to live for now."

"I am strongly against it because the people who are in control of it are not worth trusting the whole world in their hands!

It's much too much power for one person to hold."

"I think that, unless we do something about nuclear weapons the world and the human race may not have much time left (corny, huh?)."

"It gives me a pretty dim view of the world and mankind but it hasn't really influenced me."

"Everything has to be looked at on two levels: The world with the threat of ending soon, and life with future, etc. The former has to be blocked out for everyday functioning because very few people can find justification for living otherwise. But [it] is always there—on a much larger scale than possibilities of individual deaths, car accidents, etc.—even though the result to me personally would be the same."

"Yes, probably a little. It makes you wonder about how anyone could even dare to hurt others so badly."

"Quite definitely, I believe that we should try to save ourselves; any form of suicide alters the future. It would end our race."

"I sincerely hope that we stay on good terms with the USSR. I hope they never consider a nuclear war."

"I feel our growth is speeding up and if we don't slow down then we're going to die. These advances are too quick and they seem to be taking over our world."

"I feel that everyone's views of the world and ideas of the future have changed somewhat. I feel that the future is very unsettled and a nuclear war could destroy the world in a short time."

"I think that a nuclear war, which could break out in a relatively short period of time in the far future, could nearly destroy the world."

"In a way it has. It has shown me how stupid some adults can be. If they know it could easily kill them I have no idea why they support it. Once in a while it makes me start to think that the end of my time in life may not be as far off as I would like it to be."

"Yes, I feel if men keep going on with experiments they are bound to make one mistake that could mean the end of a lot of surrounding cities and if severe enough the end of what we know today as the world."

The questionnaires showed that these adolescents are deeply disturbed by the threat of nuclear war, have doubt about the future and about their own survival. There is a revelation in these responses of the experience of fear and menace. There is also cynicism, sadness, bitterness and a sense of helplessness. They feel unprotected. Some have doubts about planning families or are unable to think ahead in any long-term sense.

We may be seeing that growing up in a world dominated by the threat of imminent nuclear destruction is having an impact on the structure of personality itself. It is difficult, however, to separate the impact of the threat of nuclear war from other factors in contemporary culture, such as the relentless confrontation of adolescents by the mass media with a deluge of social and political problems which their parents' generation seems helpless to change.

It seems that these young people are growing up without the ability to form stable ideals, or the sense of continuity upon which the development of stable personality structure and the formation of serviceable ideals depend. We may find we are raising generations of young people without a basis for making long-term commitments, who are given over, of necessity, to doctrines of impulsiveness and immediacy in their personal relationships or choice of behaviors and activity. At the very least these young people need an opportunity to learn about and participate in decisions on matters which affect their lives so critically.

The experience of powerlessness of children and adolescents, the sense they have that matters are out of control, is not different from the way most adults feel in relation to the nuclear arms race. Little can be done to help our young people unless adults address the apathy and helplessness that we experience in relation to the arms race and the threat of nuclear war.

It is my contention that the madness of the arms race is not primarily in individuals but in the context of the problem. There are individuals, especially in the two superpowers, who bear responsibility for the arms race, but policy-makers and strategists seem to be caught up in a structure, a system, in which the interlocking parts activate one another, but which no one controls. There is a state of mind, a mental "set," which accompanies this system.

As an American I am part of the context of the arms race, but, insofar as I have not taken part directly in playing the nuclear arms game, I remain in ignorance of what it is like to live in the middle of it. This ignorance may be offset somewhat by the advantage of being able to look at the system, at least in part, from the outside and to ask a few questions about it.

The system, or box, contains the leaders and citizens of the United States, the Soviet Union, the other nations who have the bomb or may get it, and terrorists who may potentially get hold of nuclear arms. Inside the box it seems to be a world in which reality and imagination, psychology and politics, both domestic and international, are intertwined. It is a world in which the worst thing one can imagine an adversary might do can dictate major policy decisions and underlying assumptions which are not questioned.

Often those who come into positions of responsibility from outside of the federal government see the nuclear arms race initially as unnecessary, believing that many fewer weapons would suffice. But they end up trapped in the system, caught in the escalation game. This happens to presidents as well as many academics, including those who have or still do regard themselves as against war or violence. Once caught up in the structure of the arms race, all this appears to change, and each individual gets caught in the unending process of seeking security through more arms, in strategies of besting the other side through superior weapons and weapons systems.

How does this happen? What is there in the climate of decision and policy-making in the nuclear weapons field that absorbs men and women, although not all of them, into the dangerous process of the escalating arms race?

Before looking at the U.S.-Soviet relationship, it might be useful to consider what seems to have happened among American strategists and analysts who address questions of military security. When one considers that all share a common dedication to security, to deterrence and to the desire to avert nuclear disaster, it is remarkable to discover that there are two camps of almost diametrically opposite persuasion. To identify the two groups descriptively I will refer to them as the "unthinkables" and the "thinkables," rather than using such names as "liberals" and "conservatives," or "hawks" and "doves," whose meanings are unclear or value-laden. Both sides are aware that a nuclear war could in fact happen, that is, can think about the possibility of it, so it is not in this sense that I am using the terms. Rather, I would include in the unthinkable group those who believe that a nuclear war, once begun, is likely to create a disaster of such magnitude that it is not meaningful to plan in terms of its actual occurrence. The thinkables, in contrast, believe that one should plan in terms of nuclear war actually occurring, and even for its aftermath.

Colin Gray, one of the representatives of the thinkables, predicted that:

> "If one designed a simple questionnaire containing, say 10 'litmus paper-type' test questions of an either/or character, and submitted this questionnaire to 100 members of the U.S. national security community, inside and outside of government, there would be little cross voting by individuals between 'liberal' and 'conservative' replies."[1]

After reviewing some of the recent writings of the two groups I am convinced that Gray is right. From these articles I have extracted my own 10 test questions and representative responses which are presented in the table (pages 30–31).

What can be said of such a gulf in thinking? Surely it reflects a different experience of reality, or that the two groups perceive different realities to exist. I suspect that no amount of reasoned argument would persuade one side of the rightness of the other's position. At one level the differences might be reduced to the conviction on the part of the unthinkables that the thinkables are unrealistic about the realities and dangers of nuclear war, whereas the thinkables believe that the unthinkables insufficiently appreciate the menace of the Soviet Union. But there is more to the problem than this.

I will not attempt to interpret the motivation of the two groups, but rather the context which gives rise to the difference. My suspicion is that what underlies the difference derives from the atmosphere of unbearable terror within which the participants in the arms control community live. It is a world in which individuals who have to be, or choose to be, active must bear some portion of responsibility for the safety or annihilation of the human race. This terrifying context brings forth mental sets or structures deep within the human psyche that need to be understood.

The unthinkables seem more willing to experience directly, or *hold* emotionally, the reality of the nuclear danger, although at times they seem not to appreciate the menace of Soviet political and military strategy as the thinkables perceive it. The thinkables appear unable to experience, or have found a way not to experience, the terror of the nuclear reality itself. Or, stated differently, the thinkables have found a way to avoid its terror by reverting to older, more comfortable and familiar, war-making thinking. They may quote the Prussian general and theorist of war Karl von Clausewitz and his "winning" strategies, and write about "fighting" a nuclear war in response, they say, to similar thinking on the Soviet side. One of the most striking examples of such thinking is the following:

"Any American president should know that the only kind of war his country can fight, and fight very well, is one where

The 'thinkables' and the 'unthinkables'

Among U.S. military strategists and analysts who share a common dedication to security, deterrence and the desire to avert nuclear disaster, two camps exist of almost diametrically opposite persuasion. From some of the recent writings of the two camps—the 'thinkables' and the 'unthinkables'—ten test questions and representative responses were extracted.

'Thinkables' believe that one should plan in terms of nuclear war actually occurring, and even for its aftermath.

When the Soviets say that nuclear war is unthinkable, or mutually suicidal, it is for U.S. consumption, to lull U.S. policy makers and the public, to strengthen the anti-military in the United States.

Competition is secondary to security and to strategic necessity. We compete so that we can get the advantage.

The Soviet Union is making significant advances in civil defense and dispersion of industry so that their population and industry will no longer be hostage. This is altering the strategic balance.

We must base strategy on how it will be, or be expected to be, after the attack, as it is upon this that Soviet strategy is based, that is, on emerging the winner—in extreme form that we can emerge as winners.

The Soviets plan coldly, militarily, and are more inured to death due to their World War II experience, to the losses of millions in purges, and are willing to sacrifice large portions of their population if it is necessary to win.

That is an outrageous position. No one would want nuclear war. It is awful, but the Soviet danger is real and a strategic approach is necessary.

The Soviet Union is ideologically committed to world domination and will press the United States to the brink. Only an effective nuclear strategy can deter this.

We need to plan for limited nuclear wars and for limiting them once started.

We should strive for superiority. The Soviet Union is pursuing a winning approach, and we must defend ourselves by developing one also.

Nuclear war is possible. We should plan strategically for it.

'Unthinkables' believe that a nuclear war, once begun, is likely to create a disaster of such magnitude that it is not meaningful to plan in terms of its actual occurrence.

When the Soviets say that nuclear war is unthinkable, or mutually suicidal, we have to take seriously that they mean it.

Competition and construction of new weapons systems produces comparable responses on the Soviet side and escalates dangerously the arms race.

No known technology can effectively defend against nuclear attack (ABM, shelters). No meaningful civil defense is possible.

It makes little difference how things will be "after" the attack, as there will be no viable civilization remaining for either us or the Soviet Union. There will be no winners and the living will envy the dead.

Many Soviets are as sensitive as we are to the potential disaster to both of our peoples of a nuclear exchange.

The thinkables are less sensitive to human horrors, and can contemplate mass destruction casually, even to the point of "dehumanization" of other nations.

The Soviet Union may in fact be expansionist, but the Russians are as terrified of nuclear war as are we. They will press their interests, but not risk nuclear war.

The outcome of a nuclear war, no matter how "limited," could not be predicted. It is unlikely that any effective command structure would survive.

Nuclear superiority is meaningless and impossible. The notion of "winning" strategic approaches is outmoded, dangerous and irrelevant to nuclear conflict.

Nuclear war must be prevented at all costs.

there is a clear concept of victory—analogically, the marines raising the flag on Mt. Suribachi is the way in which a president should think of American wars being terminated."[2]

More typical are statements like:

"The question of military or political victory if deterrence fails would depend upon the net surviving destructive capacity of the two sides after the initial counterforce exchanges."[3]

It needs to be recognized that the mentality of winning, or of achieving strategic superiority (as if nuclear exchanges were really war at all in any past sense rather than spelling certain annihilation of the protagonists and of large remaining portions of the human race) is useful to its possessor in several ways. It can, at least in an immediate sense, be politically supported and supportive for national leaders. It seems to offer a reasoned response to the Soviet Union's menace. It can offset through denial and distortion the actuality of annihilation. It seems to deal effectively with the sense of helplessness or powerlessness, particularly among a people accustomed to having its own way and to experiencing a sense of power, dominance, effectiveness and success in international dealings.

I suspect that the great interest in the last few years in disaster films about air crashes, earthquakes, tidal waves, and fires in tall buildings grows out of an unconscious need to displace the larger terror contained in the threat of nuclear disaster and annihilation to a smaller, more finite, comprehendible and manageable catastrophe.

If there are special dangers in the mentality of victory, of superiority, of either/or, or of "you win/I lose," then what alternative is there? How are we to get out of the box?

One possibility is to find a way to come to terms with the full reality of nuclear arms, to acknowledge their actual menace. This would be a step toward the realization that, unlike in the case of previous wars, it is possible that security will not be provided by the development of more arms. The latter may, instead, create more danger. To move in such a direction requires a paradigmatic shift of thinking. It means living for a time with a particular kind of terror, experiencing our helplessness in the face of it, acknowledging fully the menace of the arms race and of our responsibility for the creation of this terror.

There are few willing so far to bear such fear, to experience the despair that accompanies the reality of confronting the arms race as it is. Members of a limited number of organizations and private citizens, including some arms strategists and policy-makers, have been willing to begin to move in this direction. Group support is essential.

The terror of these weapons, with their awesome destructiveness, proliferating under adversarial conditions that cause death on a scale previously unknown to be but a thin thread away creates a context that is beyond human endurance and gives rise to maladaptive positions and archaic ways of thinking that sacrifice one aspect or another of reality. Yet our survival offers no choice but to recognize this condition in which we exist, experience the reality of our own terror and to perceive and appreciate the positions of various human groups that are held within the context of this actuality. Only then can we move away from the win/lose, we or they mentality, which is likely to spell the mutual destruction of the United States and the Soviet Union. Only then can we move to the mental and emotional context of win/win, lose/lose, of we *and* they, that the development of nuclear weapons requires.

This shift in thinking needs to be accompanied by a comparable shift on the Soviet side, although such a change beginning on one side could generate a similar change on the other, just as escalation or the creation of fear by one side inevitably produces a response in kind on the other.

I have little to say about approaching the Soviet Union in actual security discussions. Analysts of Soviet intentions with regard to nuclear weapons seem to come away from their studies with previous convictions confirmed, selecting data that grow out of their own fear or preconceptions. The Soviets on their part find confirmation of their fears and distrust in the statements and actions of our policy-makers. Each move on one side generates further terror and escalation on the other and brings another turn on the thermonuclear ratchet wheel.

In the nuclear arms race the United States and the Soviet Union in effect create one another. What one side regards as protection of its legitimate interests, the other perceives as a menacing threat against its homeland. The Soviets, like ourselves, have military strategists who seek security through more weapons and write scenarios of superiority and victory in nuclear arms and nuclear war. But they also have members of their security community who have addressed the horror of the arms race and faced the reality that there can be no victory once a nuclear exchange begins. To win means to lose everything.

There seem to be in both countries individuals at various levels of decision-making who know that no security lies in the creation of new weapons systems. On the contrary, our mutual survival depends on an unceasing effort to reduce tensions through words and acts that create trust, not terror. But there is a danger that the voices of these individuals will be drowned by the strident claims of those who are trapped in the box—and escape the terror and the reality of the annihilation which faces us because they continue to believe in the false assumptive system of the arms race.

There can, in my opinion, be no differences between the United States and the Soviet Union which warrant the level of risk of nuclear annihilation we are now creating for each other and the rest of humanity. As we and the Soviet Union mutually create distrust and fear, we can also create each other's trust and security.

1. Colin S. Gray, "Nuclear Strategy: The Case for a Theory of Victory," *International Security,* 4 (Summer 1979), p. 58, n. 9.

2. Gray, "Nuclear Strategy," p. 71, n. 42.

3. Paul H. Nitze, "Assuring Strategic Stability in an Era of Detente," *Foreign Affairs,* 54 (January 1976), p. 213.

3 Buying death with taxes: impact of arms race on health care

VICTOR W. SIDEL

Physicians for Social Responsibility has concentrated its efforts on making the medical consequences of war widely known. But, as the Council for a Livable World and other groups have pointed out, military expenditures are themselves destructive of human life, even if the weapons we stockpile are never used. The diversion of a large part of the world's resources to preparation for war leaves far less available for health services and for other efforts that would improve the duration and the quality of life of the world's peoples.

We need not wait for the ultimate horror—the "last epidemic" —that nuclear war represents. Preventable endemic and epidemic disease, hunger, misery and premature death surround us now, in the midst of a potentially productive and bountiful world. Much of the world's illness, particularly in developing countries, could be prevented or ameliorated by the redirection of a fraction of the resources that the world diverts to arms.

Ruth Leger Sivard, in *World Military and Social Expenditures 1980,* has analyzed world arms expenditures in detail.[1] Military expenditures in her compilation are defined, as they are in the official budgets of NATO countries, as "current and capital expenditures to meet the needs of the armed forces. They include military assistance to foreign countries and the military components of nuclear, space, and research and development programs."

Military expenditures in Sivard's analysis do not, however, include such expenditures as veterans' benefits, interest on war debts, spending for civil defense, and outlays for strategic industrial stockpiling. They also omit extrabudgetary economic costs, such as tax exemptions accorded military properties. For estimates of Soviet and Warsaw Pact military expenditures Sivard has used a complex procedure based on a variety of sources. These estimates, for all countries, tend to underestimate seriously the burden on the economy.

It has been estimated that since World War II military expenditures for all the countries of the world amount to about $8 trillion (in current dollars).[2] (A summary of the expenditures, according to Sivard's figures for 1960 to 1978, is shown in Table 1.) World military expenditures in 1980, based on these compilations—and therefore markedly underestimating total economic impact—come to approximately $500 billion per year (equivalent to about $1.4 billion per day or $1 million per minute).[3]

It is possible to be much more precise about U.S. military expenditures. Those included in the formal U.S. budget, under the heading "National Defense," encompass Department of Defense military activities, Department of Energy nuclear weapons and naval reactor activities, and other "defense-related" activities (Table 2).[4] The $136 billion U.S. military expenditure in Fiscal Year 1980 was equivalent to about $400 million per day or $300,000 per minute. It represents approximately 6 percent of the U.S. gross national product for 1980.

Table 1

World Military Expenditures, 1960 to 1978

	Military expenditures (1977 $ in billions)	*Gross national product (1977 $ in billions)*	*Military expenditures as percentage of GNP (percent)*
NATO countries	3,137	55,952	5.6
Warsaw Pact countries	1,401	16,410	8.5
Other developed countries	263	14,182	1.9
Developing countries	935	21,570	4.3

Seymour Melman expanded the estimate of total U.S. military expenditures by including the continuing costs of past wars and military programs and of other programs justified on grounds of national defense.[5] The total expenditures for 1980 amounted to $194 billion. This represents approximately 9 percent of the U.S. gross national product—for that year equivalent to approximately $900 per person or $3,500 for a family of four in the United States that year.

From 1979 to 1980, military spending in the United States increased by 14 percent, an increase faster than the rate of inflation. Nonmilitary spending by the federal government increased by only 5 percent overall, an actual cut for most social programs, given the impact of inflation. The Carter budget for fiscal 1981 proposed outlays 3.3 percent higher in real terms, after inflation, than in fiscal 1980. The Reagan budget increases that figure and proposes that the military budget increase from $146 billion in 1980 (with $136 billion in actual outlays) to $374 billion in 1986. The $226 billion Pentagon budget (with $189 billion in outlays)

Table 2

U.S. Expenditures in Budget Category "National Defense"

Fiscal year	*Outlays* (*current $ in billions*)	*Increase over previous year* (*current $ in billions*)	(*percent*)
1977	97.50[a]	--	--
1978	105.20[a]	7.70	7.9
1979	117.70[a]	12.50	11.9
1980	135.85[a]	18.15	13.4
1981	162.10[b]	26.25	19.3
1982	188.85[b]	26.75	16.3
1983	226.05[b]	37.20	16.5
1984	255.60[b]	29.55	13.1

[a]Actual
[b]Reagan administration request (March 10, 1981)

proposed for 1982 represents the largest U.S. peacetime increase in the military budget in history.* Under this proposal, U.S. military expenditures would come to some $1.3 trillion over the next five years, compared to the approximately $2 trillion (in current dollars) spent on arms during the 35 years from 1946 to 1980.

Other countries have also markedly expanded their military spending.[6] The level and trend of military spending in the Soviet Union are a matter of disagreement, but the U.S. Department of Defense estimates that Soviet military spending has risen at a rate of 3 to 5 percent annually in the 1970s. The general upward trend outside NATO, the Warsaw Pact and China has been of the order of 7 to 8 percent a year from 1970 to 1979. The increase in many of the poorer countries considerably outstripped inflation. In South America, for example, military spending has increased about 5 percent a year in real terms.

Most countries are said to be planning increases in their military expenditures above the inflation rate over the next few years; NATO asked its member countries in 1978 for a 3 percent real increase each year in military spending.[7] The Soviet Union and the Warsaw Pact countries, as well as many Third World countries, apparently also intend to continue to increase in real terms their annual outlays on arms.

What are the effects of the arms race on the economy and on health services?

Diminished productivity. The massive arms investment in the United States and in a number of other countries has diverted capital from the modernization of productive capacity for the civilian economy. Countries such as Japan and, until recently, West Germany, which spent far less on arms, have far surpassed the United States in their rate of growth in manufacturing productivity (see figure).[8] Expansion of health and other human services

---

*Military budget authority is the amount of money that can be contracted to be spent in future years; military outlays is the amount to be spent in a specific fiscal year.

depends largely on an expanding "economic pie." If the pie is not increasing, it is harder to convince people, and their political representatives, to increase the amount spent on publicly-funded services for the underserved.

Increased inflation rate. Expenditures for arms production are much more inflationary than are expenditures for consumer goods and services because they pump money into the economy without increasing the supply of goods and services. An editorial entitled "Burning $1 Trillion" in the January 22, 1980 Wall Street Journal said that "Defense spending is the worst kind of govern-

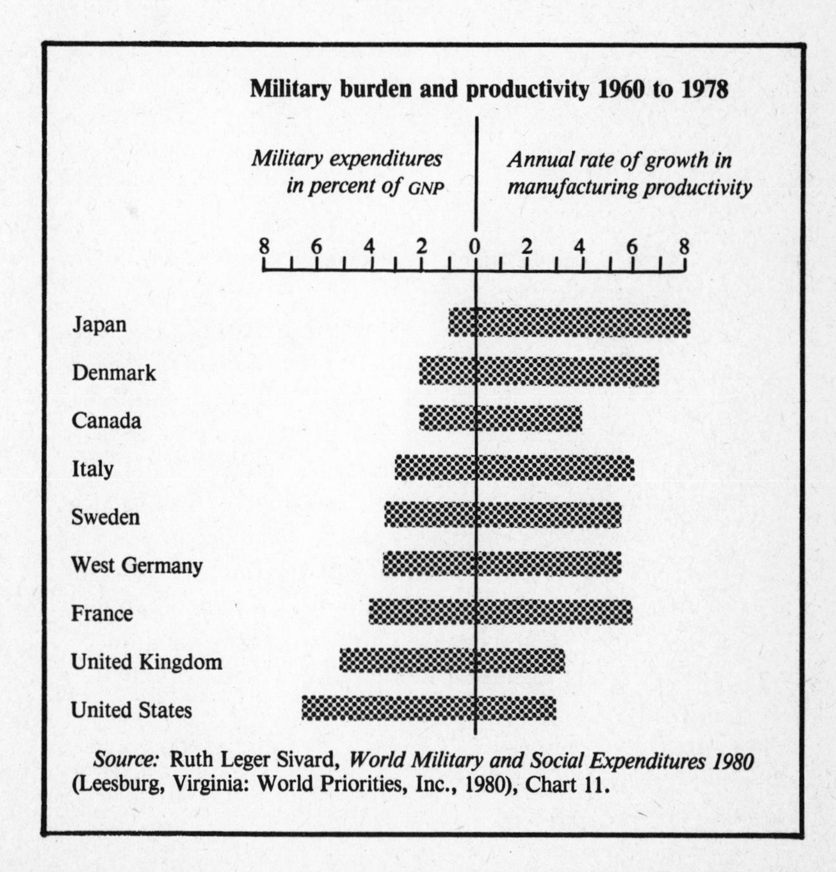

Source: Ruth Leger Sivard, *World Military and Social Expenditures 1980* (Leesburg, Virginia: World Priorities, Inc., 1980), Chart 11.

ment outlay," because it not only increases income without increasing the supply of goods that consumers can buy but it also "eats up materials and other resources that otherwise could be used to produce consumer goods." Inflation is especially hard on the poor and those on fixed incomes; by decreasing disposable income it reduces personal funds available for health care and by decreasing purchasing power it reduces in real terms public funds available for health care.

Increased unemployment. Expenditures for military production create far fewer jobs than expenditures for human services. When the United States spends $1 billion on arms, for example, it creates some 40,000 jobs; the same amount spent on nursing creates some 80,000 jobs. The International Association of Machinists and Aerospace Workers estimated in 1979 that $78 billion of military spending cost 3.25 million civilian industrial and service jobs.[9] Unemployment also generally hurts the poor more than the affluent. Furthermore, since many health insurance programs are tied to employment, unemployment often terminates health insurance coverage.

Diversion of other resources from health programs. The arms race diverts able people from health and other human services to military functions, a problem that exists in all countries, but is worse in poor ones. The arms race also diverts technical capacity from medical and other civilian technology to military technology. And finally, there is a drain of skilled researchers from medical and other quality-of-life-oriented research to military research. Of $30.7 billion U.S. federal research and development spending in 1980, $15.1 billion (49 percent) was spent on the military and $3.7 billion (12 percent) on health. Of the world's research physicists and engineering scientists, more than half are working only for the military.[10]

The American Federation of Clinical Research has analyzed the impact of the Reagan administration budget proposals for fiscal 1981 to 1986 on the budget for the National Institutes of Health

(Table 3). Because the annual increases are considerably less than the expected inflation rate, if these proposals are adopted the Institutes will *lose* 25 percent of their 1980 buying power by 1986.[11]

Diversion of public monies from health programs. Most directly, of course, the arms race shifts tax revenues away from human services. This is true not only at the federal level, which provides 28 percent of U.S. health expenditures. Because additional service burdens are put upon states and localities, which now provide 11 percent, the arms race reduces their ability to fund health services as well. Some of the areas in greatest need lose the most. Using the Carter administration budget for fiscal 1981 to 1985 (which is of course less than the Reagan budget), James R. Anderson estimates that the New York metropolitan area will send $56 billion to the Pentagon over this five-year period; Chicago will send $42 billion; Detroit $25 billion; Boston $20 billion; Newark $12 billion; and Cleveland $11 billion.[12] All of these cities are markedly

Table 3

Budget for the U.S. National Institutes of Health

Fiscal year	*Outlays (current $ in billions)*	*Increase over previous year (percent)*	*Buying power in 1980 $ (deflated $ in billions)*[a]	*Decrease in buying power compared to 1980 (percent)*
1980	3.44[b]	--	--	--
1981	3.52[c]	2.3	3.24	8
1982	3.76[c]	6.8	3.18	9
1983	3.88[c]	3.2	3.02	13
1984	4.00[c]	3.1	2.87	17
1985	4.13[c]	3.3	2.72	21
1986	4.27[c]	3.4	2.59	25

[a]Assuming 8 percent inflation rate
[b]Actual
[c]Reagan administration request

reducing essential human services, and several are already close to bankruptcy.

The relationship between monies spent on the arms race and monies spent on human services was dramatically indicated by Edward F. Snyder of the Friends Committee on National Legislation in his testimony on the 1979 federal budget before the House Budget Comittee:

"Six surveillance and command planes added for an additional cost of $120 million; Headstart programs (pre-school training) cut $20 million. . . . One submarine tender to support nuclear attack submarines added for an additional $262.5 million; maternal and child health services cut $112 million. . . . An additional $89 million for the Army XM-1 tank program; mental health centers cut $88 million. . . . An additional $141 million for A-10 attack aircraft; cancer research cut $55 million"[13]

Although the United States ranks relatively high in health status among the nations of the world—although not as high as its relative wealth should allow—health and health services are grossly maldistributed.[14]

Health problems of the poor and of racial minorities are much worse than those of the affluent and of whites. Overall U.S. mortality rates are declining, but the mortality of social minorities is still one-third greater than that of whites, and higher rates are also found among the poor and uneducated. Death rates in every year of life are higher for poor children and highest for poor black children. Infant mortality among racial minorities is almost twice as high as that among whites. Life expectancy at birth for members of racial minorities is over five years shorter than for whites. Non-whites have three times the maternal mortality of whites, nor has this difference been reduced significantly in the past 25 years. From 1950 to 1975 the age-adjusted cancer death rate for the white U.S. population increased 4 percent; for non-whites the increase was 20 percent. In a low-income Chicano community in Los Angeles, children were found to have four times as much amoebic dysentery as the national average; twice as much measles, mumps and tuberculosis; 1.4 times as much hepatitis.[15]

In New York City, where I work, the infant mortality rate in the South Bronx in 1970 to 1972 was 19 percent greater than for the city as a whole; by 1977 to 1979 the South Bronx rate was 33 percent higher than for the city as a whole. In New York City new cases of tuberculosis increased by almost 20 percent from 1978 to 1979, the first year in the past 30 that the number of new cases increased over the previous year. And other statistics are just as appalling.

Health services for the poor and for racial minorities, most of which must be publicly funded, are much less adequate than those for the affluent and for whites. In many cities, resources for public health services and for municipal hospitals are grossly inadequate. Funds for these programs, in real terms, are diminishing in many areas. Public hospitals have recently been closed in Philadelphia, St. Louis, and New York City, and are in serious trouble in other cities such as Chicago and Atlanta.[16]

As a result of the destruction of public services, people in inner cities as well as rural areas are becoming even more underserved. The budget cuts proposed by the Reagan administration will specifically reduce health services for the poor through reduction in transfer payments, in entitlement programs such as Medicaid, and in categorical programs such as rat control, lead paint and lead poisoning screening and control, and community and migrant health centers. The proposed cuts in the National Health Service Corps will destroy one of the few effective mechanisms for placing health care personnel in medically-underserved urban and rural areas. Relatively small amounts subtracted from the U.S. military budget could make an enormous difference. For example, the entire annual budget of $20 million for the U.S. Public Health

Service Hospital in Seattle, scheduled to close in 1982, is less than two-thirds the cost of one AH-64 military helicopter.

The discontinuation of the National Health Service Corps would over time decimate the 22 community clinics in Seattle and would remove the five health professionals who provide health services at the King County Jail. Other medically underserved urban and rural communities in the United States are even more dependent on the Corps. The budget cuts for the National Health Service Corps, nationwide, for the next two years are $17 million —the equivalent of the cost in fiscal 1982 of seven XM-1 main battle tanks, or one hour of U.S. military expenditures. The Reagan administration projects cuts of over $1 billion in federal Medicaid funds and over $1 billion in U.S. Public Health Service programs in 1982; these cuts could be more than restored by eliminating the $2.4 billion budget for *development* of the MX missile program in 1982. Just a part of the budget *increase* for arms proposed by the Reagan administration ($53 billion by 1982) could vastly improve health status and health services in the United States. Cutting the already grossly inflated military budget would, of course, help even more.

The health of people in poor countries is generally much worse than that of people in affluent countries. Two billion of the world's people do not have access to a dependable, sanitary supply of water; in some countries, less than 5 percent of the population has access to safe water. Water-related diseases kill approximately 10 million people every year. Life expectancy at birth is 30 years shorter in Africa than in Europe. Infant mortality rates overall are six times as high in developing countries as in technologically-developed countries. Among children under 15, there are over 12 million annual deaths from diseases that could be prevented by immunization.

Health services in poor countries are far less developed than in affluent ones; in some they are almost non-existent. In a number of developing countries there are fewer than 10 physicians per

100,000 population, compared to 100 or more per 100,000 in most developed countries. In a number of developing countries there are fewer than 50 hospital beds per 10,000 population, compared to 500 or more per 10,000 in most developed countries. Health services in most of the poorer countries are concentrated in the major cities, with little available to poor rural populations. Many poor countries spend less than five dollars annually per capita on health services.

Aid from developed to developing countries is diminishing. The generally-agreed-upon international target for such aid is 0.7 percent of the developed countries' gross national product. Actual aid from countries affiliated with the Organization for European Cooperation and Development, and from the United States, is less than half that, while from the Soviet Union and the Eastern European countries it is less than one-tenth.

Even as military expenditures have risen, there have been widespread cuts in the real value of development aid to the Third World. And many developing countries purchase large quantities of arms from developed countries, transferring resources from the poor countries to the rich ones. Two of every three countries in the world spend more public monies on arms than on health services.[17]

Redirection of relatively small amounts of the resources devoted to the world's arms race could make an enormous difference. The total cost of the program that eradicated smallpox from the Earth is less than the cost of six hours of the world arms race. The entire cost of the malaria-control program of the World Health Organization is less than the cost of one day of world military expenditures.[18] Relatively simple and inexpensive technologies to help people provide their own safe water, immunization and adequate nutrition—which could save tens of millions of lives annually and improve the quality of life for hundreds of millions—are well-known. All that is needed is a diversion of some of the world's resources squandered on arms.

A recent statement by the Public Health Association of New York City concluded:

"The health of the people of New York City is actively endangered by the already imposed cuts and by the threatened cuts in funding for health care services and for medical care services. To express ourselves in clear language, so there is no misunderstanding: We are talking about dead babies whose deaths can be prevented; we are talking about sick children and adults whose illnesses can be prevented; we are talking about misery for older people whose misery can be prevented. We are speaking of these unspeakable things in a wealthy country and in a wealthy state, whose people deserve better. The malignant neglect of federal, state and local governments is literally killing people now and will kill, and destroy the lives of, many more in the future. We urge a massive infusion of federal and state funds to restore and rebuild services now . . . before the consequences of their breakdown demonstrate in even more tragic and dramatic ways the human and economic costs of this neglect."[19]

And an editorial in the prestigious British medical journal, *The Lancet*, began: "The medical profession must never neglect its responsibility to protest at the grim paradox between the world's enormous and mounting military expenditures and the comparatively meagre efforts devoted to the relief of poverty, malnutrition, and disease."[20]

The responsibility, in short, is ours.

1. Ruth Leger Sivard, *World Military and Social Expenditures 1980* (Leesburg, Virginia: World Priorities, Inc. 1980).

2. Frank Barnaby, "Global Militarization," *Proceedings of the Medical Association for the Prevention of War,* 3 (March 1980), p. 132.

3. Bernard Lown and others, "The Nuclear Arms Race and the Physician," *New England Journal of Medicine,* 304 (March 19, 1981), p. 726.

4. U.S. House of Representatives, Committee on the Budget, *Chairman's Recommendations for the First Concurrent Resolution on the Fiscal Year 1982 Budget* (Washington, D.C.: U.S. Government Printing Office, April 6, 1981).

5. Seymour Melman, "On the Social Cost of U.S. Militarism," *American Journal of Public Health,* 70 (September 1980), p. 953.

6. *Armaments or Disarmament?* (Stockholm: Stockholm International Peace Research Institute, 1980).

7. Frank Barnaby, "A 3% Rise in Military Spending," *Lancet,* 1 (June 7, 1980), p. 1231.

8. Sivard, *World Military*, p. 15.

9. "Reagan Would Spend Us Into the Grave," *In These Times,* 5 (April 8, 1981), p. 14.

10. Frank Barnaby, "A 3% Rise."

11. Anthony S. Fauci, American Federation for Clinical Research, memorandum (April, 1981).

12. James R. Anderson, "Bankrupting America: President Carter's Military Budget," *International Journal of Health Services,* 10 (1980), p. 581.

13. Vincente Navarro, "The Social Costs of National Security or Insecurity," *American Journal of Public Health,* 70 (September 1980), p. 961.

14. Victor W. Sidel and Ruth Sidel, *A Healthy State: An International Perspective on the Crisis in U.S. Medical Care* (New York: Pantheon Books, 1978).

15. H. Jack Geiger, "Small Futures, Sick Futures, Short Futures: Inequity and Irrelevance in U.S. Health Care Strategies," *Working for a Healthier America,* Walter J. McNerney, ed. (Cambridge, Massachusetts: Ballinger, 1980).

16. E. Richard Brown, "Public Hospitals in Crisis: Their Problems and Their Options," *Health Activists' Digest,* 2 (Fall 1980), p. 3.

17. Sivard, *World Military.*

18. Lown, "Nuclear Arms Race."

19. Victor W. Sidel, "Statement on Behalf of the Public Health Association of New York City," Statewide Conference on the Public Health Crisis, Albany, N.Y. (May 29, 1980).

20. "Threat of Nuclear War," *Lancet,* 2 (Nov. 15, 1980), p. 1061.

4 Economics of the arms race —and after

JOHN KENNETH GALBRAITH

Anyone who wishes to deal responsibly with the nuclear arms race between the Soviet Union and the United States and with its consequences must begin by conceding the diversity of the motivating forces that are involved.

There is, first of all, the technological trap. Each power develops the weapons which make obsolete those of the other; anticipating this, each strives to develop those that protect it from that obsolescence and provide an advantage instead. The resulting interaction has a technological dynamic of its own.

This dynamic is then sustained by economic, bureaucratic and military interest. Public and private employment, bureaucratic expansion by the Pentagon and weapons firms, personal prestige and corporate earnings are all generated by the arms race. This is the influence of which Dwight D. Eisenhower in his best-remembered and most eloquent statement warned a full 20 years ago. We must be on guard "against this acquisition of unwarranted influence, whether sought or unsought, by the military-industrial complex. The potential for the disastrous rise of misplaced power exists and will persist. . . . We should take nothing for granted."

There is a certain reluctance in this polite and cautious age to speak of the financial interest in the arms race. Can anything so dangerous, so catastrophic, be motivated by financial interest or personal gain? But the financial analysts and like scholars are not so inhibited; they are currently and eagerly telling their customers and clients of the corporate prospects from the coming increase in the arms budget.

It is a measure of the force of pecuniary and bureaucratic interest that it can so capture the minds of the people involved that they do not themselves reflect on ultimate consequences. A curtain is lowered over the future. Sufficient is the dollar today; let there be no thought that it means death for one's self and one's children tomorrow.

We must suppose that other and counterpart forces operate in the Soviet Union. It is not my instinct to argue that the dynamism in this race is sustained from one side. It is, however, to those on one's own side that one speaks.

There is on our side a final, more general, and yet more influential force sustaining the arms race. That is the belief that we are defending an economic, political and social system. It is by the proliferation of arms and the pursuit of ever more arcane and dangerous weapons technology that we protect free institutions, free enterprise, capitalism, the American way of life. Our institutions are under assault from socialism, communism. It is by a large and growing commitment to weaponry, at whatever cost or danger, that we protect them. It is this last motivation that I shall especially address.

The arms race as it now proceeds does not strengthen free institutions or free enterprise or however we denote our economic and social system. On the contrary, it is gravely weakening that system. And if or when, in some moment of error, anger or panic, this race goes out of control—if there is a nuclear exchange, large or as some now imagine limited—what is called free enterprise or capitalism will not survive. Nor will free institutions. All will be shattered beyond recovery. So, equally, will be what is now called communism. Capitalism, socialism and communism are all sophisticated social forms relevant only to the advanced world as it has now developed. None would have existence or relevance in the wreckage and ashes and among the exiguous survivors of a post-nuclear world. This is not a matter of easy rhetoric; involved are hard facts which no one after serious thought can escape.

But it is a thought that a great many otherwise very sensitive and sensible people do try to escape. The mind resists involvement with horror as, in a normal person, it resists preoccupation with death. And in consequence we leave the issue of nuclear arms, their control and their consequences to the men who make horror their everyday occupation. It is a reckless, even fatal, delegation of power.

The economic effects of the arms race and its adverse effect on the economy are not matters of any great subtlety. Nor are they such as to provoke any great argument between economists or other informed observers. And there is strong empirical verification of the result.

Expenditure for arms, like any public expenditure, is at cost to other use. And notably it is at cost to private capital use. This is a proposition accepted by economists of liberal instinct; it is even more explicit in the speech and policy of conservative economists. It is central in the general thought of the economists of the present Administration. The freeing of revenues now used for public purposes is deemed vital for private capital improvement—for the revitalization of the American industrial plant. The point, however, has not been stressed as regards our military outlays.

On occasion in recent years we have heard reference to the way in which military expenditures have been drawing resources from other public needs—what even a small share of the money going to the Pentagon would do to arrest the decline of our urban centers and services. Or to provide employment for the young in the big cities. We have heard much less—there has been something approaching a conspiracy of silence or neglect—of the way these expenditures have contributed to our industrial decline.

That contribution has been both general and specific. During the 1970s we spent annually around $100 billion on our defense establishment for a total for the decade of roughly $1,000 billion in constant 1976 dollars. Capital in this magnitude can be used for arms; it can be used for private capital investment; it cannot be used for both. If an appreciable part of this outlay had gone into the improvement of our industrial plant—as it would have gone had it not been requisitioned by the government—no one can doubt that our economy would be stronger today. And from this stronger economic position would have come some of the political prestige and primacy that we enjoyed in the earlier years following World War II. It was our economic, not our military, strength on which our world position then, in largest measure, depended. Military expenditure is at cost to economic strength; and it is upon our economic strength that our world position in the past has rested.

Such is the general effect: a massive transfer of capital away from civilian industry over the years and a resulting weakness that is for all to see. The specific effect within the industrial system is even more visible. Modern military expenditure has a highly specialized impact. It concentrates itself, a less than novel point, on that narrow range of industry and the highly specialized technology that serve missile, aircraft and marine weaponry. There can be little doubt as to the stimulating effect on this industry and this technology. But this development and the associated distortion in the allocation of resources—the technical competence, capital, labor and other resources lavished on this small, specialized sector of the economic system—have been at heavy cost to the industries on which we depend for domestic or international competitive performance. As we have pressed ahead on a narrow band of industry that serves our weaponry, we have left behind, left competitively vulnerable, our steel, automobile, textile, chemical and a great range of other industries.

There is no one reason for the decay of our standard civilian industry; one does not strengthen a strong case by attributing exclusive causation. But let no one doubt that the weakness in our older industry is the counterpart of the distorted growth of that narrow band of industry that supports the arms race.

It is hardly news that in modern times our competitive position has declined steadily in relation to that of Germany and Japan. We are not less intelligent than the Germans and Japanese. Our raw material and energy base is not less good—indeed, it is far better. It is not clear that our workers are less diligent. Germany spends more per capita on its social services than we do; the Japanese do not spend much less. The difference is that the Germans and Japanese have been using their capital to replace old civilian plant and build new and better plant. We have been using our capital for industrially sterile military purposes. The comparison is striking.

Through the 1970s we used from 5 to 8 percent of our gross national product for military purposes. The Germans during this decade used between 3 and 4 percent—in most years about half as much as we did. The Japanese in these ten years devoted less than 1 percent of their gross national product annually to military use. In 1977, to take a fairly typical year, our military spending was $441 per capita; that of Germany was $252; the Japanese spent a mere $47 per capita. It was from the capital so saved and invested that a substantial share of the civilian capital investment came back which brought these countries to the industrial eminence that now so successfully challenges our own.

Again the figures are striking. Through the 1970s our investment in fixed nonmilitary and nonresidential investment ranged from 16.9 percent of gross national product to 19.0 percent. That of Germany ranged upward from 20.6 to 26.7 percent. The Japanese range in these years was from 31.0 percent to a towering 36.6 percent.* The investment in improvement of civilian plant was broadly the reciprocal of what went for weapons. Can anyone looking at these figures suppose that our military spending has been a source of industrial and economic strength? Can anyone doubt that it has been at cost to our industrial eminence and to the prestige and influence that go therewith?

*Figures are from *The Statistical Abstract of the United States and International Economic Indicators* (Dec. 1980).

Let us ask ourselves again. Have we strengthened our position in the world by accepting a decline in our civilian industry? In an age of overkill, do we win industrial strength by investing in yet more overkill? At a minimum would we not be wise, in the most conservative sense, to urge the arms control that would allow us to use more of our capital for improving the competence and excellence of our civilian industry? I do not appeal to some pacific idealism but to practical self-interest. I cannot think there can be any doubt as to the answer.

What of the aftermath of the arms race—the all-too-certain consequence if it is not brought under control? We have heard of the death, direct and enduring, that must follow a nuclear exchange. And of the peculiar horror that must follow for the survivors when physicians, nurses, hospitals and medicines are swept away with the dead. We have not sufficiently heard that the economic system on which we depend—and which, as noted, we presume to defend—will succumb no less completely than those who live in the target areas. And it will not matter greatly what those areas are.

In the most general sense our system depends on a voluntary response to pecuniary motivation—on a self-motivated decision to work for money. The one great certainty is that, after the first exchange, this motivation will no longer serve. People will not work for money if to do so means risking sickness and death and if, as a further certain consequence of catastrophe, they will not be able to spend what they earn.

To pass again from the general to the specific, we may consider transportation. On this, after only the shortest time, all life depends. In the aftermath of the nuclear exchange there will be no transportation system. Surviving truck drivers and trainmen will not accept for mere pay the risk of passing through areas of devastation or contamination or of further attack. In consequence, the transport of goods will come to a halt; after stocks of food of varying degrees of contamination are exhausted, starvation will ensue. Likewise exhausted will be stocks of medicine and fuel. Starvation and exposure will be the fate of those so unfortunate as to escape the initial blast and fire.

Perhaps, it will be supposed, compulsion can be brought to bear on those who drive the trucks and trains—a draft into the armed services or some equivalent. That would be a formidable exercise in organization under the most favorable of circumstances; it would be impossible in the general anarchy of the aftermath. And the fact that some form of compulsion is the only answer emphasizes my point: the economic system based on established motivation would be one of the first casualties of the blast.

One can pursue the consequences— and there is no need for exaggeration or hyperbole. Surviving men and women at all occupational levels will reject factory and other work in the target areas or possible target areas. Again there is no pecuniary payment that makes worthwhile the certainty of death. But, the most urgent essentials of life apart, there will be no purpose in such production. A people desperately preoccupied with remaining alive will not wish for automobiles, washing machines or toothpaste. And what might be produced cannot in any case be shipped. There will be no education; the cultural heritage of one thousand years will be gone. Let this be a thought every time you pass a university, enter an art gallery or a museum.

But no one should suppose that even the provision of essentials will be possible. Food is the obvious case in point. The production of food in the modern economy is tightly interlocked with other industry. Farms once, within our lifetimes, supplied themselves with their own power; now they are inescapably at the end of a long supply line that brings them fuel for tractors and dozens of lesser requisites for production. Transportation again becomes decisive; without it, food will be neither produced nor delivered. And the question also arises as to why farmers will struggle to provide food for others when they can get nothing in return. Again, money in whatever amount and in whatever form will cease to be an incentive. As organized compulsion would be necessary to maintain a transportation system, so it would be required to ensure the production of food and other essentials. And as this possibility is contemplated, it becomes ludicrous.

The point has been sufficiently made. The modern economic system cannot stand the shock and the associated fears from any kind, limited or unlimited, of nuclear exchange. The transportation network, as noted, that knits the various parts of this system together will be the first casualty; without transportation all parts will succumb. But transportation is a metaphor for the shattering vulnerability of our highly sophisticated, deeply interdependent system.

One returns to all who see in nuclear war a way of defending an economic and social system. This is not a relevant effort for the millions who will already be dead. But it has no greater meaning for those who survive. In the nuclear aftermath free institutions, free enterprise, capitalism and associated property and pecuniary values will all be gone. Those who say that we should accept the risk of nuclear conflict to save our system are saying, in the strongest possible terms, that we should accept its certain destruction.

A year or so ago a Pentagon spokesman, reporting on an internal study, said that the United States, in the aftermath of a nuclear exchange, could sustain a "medieval" standard of living. The statement was apt—perhaps more apt than the spokesman imagined. Under medieval economic conditions, men and women grubbed a living from the soil over a short and recurrently disease-ridden life. They had what was made with their hands; there was no manufacturing, little trade. There was nothing that could be called capitalism or free enterprise—or freedom. It was not that these had yet to be invented; it was rather that they belong to a far later and higher stage of economic development. They were born not out of ideological invention or preference but out of historical change.

There was little personal freedom in the medieval system, because it could not be afforded. Men and women were tied to the land because in the medieval economy that compelled association was the alternative to pecuniary compensation. Property rights and values were of little consequence; property is of significance only when it has value in production or in use. So it all would be in the post-nuclear world. Everything that is now known as free enterprise or the free economy would be gone. So likewise socialism. As capitalism and communism would be indistinguishable in the ashes of the target areas, they would be wholly irrelevant for those who survived.

The lesson is clear. In co-existence with the Soviet Union and the rest of the socialist or communist world, the alternative economic and social systems can survive. The arms race has a deeply damaging effect on the free enterprise economy, and we may assume that this is equally true for the socialist system. In our case the results are sadly visible—our industrial plant starved of capital, weakened in relation to our competitors, world prestige lost—all from the diversion of capital to the arms race. This already has been the tangible damage to our system, but the worst is not yet. As I have said, in the aftermath of the uncontrolled arms race, no modern economic system will survive. Both capitalism and socialism will succumb. It will be of little pleasure for even the most passionate ideologue that, in destroying the Soviet system, we destroy no less completely our own.

The only and inescapable course remains. That is, whatever our contemporary differences, to join with the Soviet Union in seeing that we do not both return—those who live—to what, not inaccurately, were called the Dark Ages. This is a matter on which our interests are wholly in common. Back of arms control, supporting the measures which ensure against devastation by calculation, miscalculation, anger or accident, lies a wholly common interest.

It is that common interest that all Americans whatever their political preference—whatever their political fears—have a concern to pursue.

We must not, regardless of our politics, continue to imagine that we can protect or save an economic system by accepting the possibility of nuclear war. In the ashes communism and capitalism, let it be said again, will be indistinguishable. They will also be indistinguishable and irrelevant in the ultra-primitive struggle for existence of those who are so unfortunate as to survive.

II CAUSES

"Both the United States and the Soviet Union have very large, diversified and ever improving arsenals of nuclear weapons that can assuredly destroy everything in both countries many times over."

—Kosta Tsipis and John David Isaacs

If they build bigger missiles to combat our anti-missiles, then we'll build still bigger anti-missiles, which will require them to build still bigger missiles.
That's terribly reassuring.

5 A clear and present danger —East

E. I. CHAZOV

This is a difficult period for mankind. Life on Earth has never been in such danger as it is now. This danger is because of the nuclear arms race and the production of weapons whose devastating power cannot be compared to that of any earlier types of weapons. The residents of Europe recall with horror World War II, which snuffed out 50 million lives, leaving behind ruined cities and villages and tragedies for millions of families. In our country 20 million people were killed and virtually every family was affected.

We know very well what war means. But what could happen to our planet in no way compares even with what humanity lived through during World War II. It has been calculated that about five megatons of various kinds of explosive substances were used during the whole of that war. To get an idea of what can happen today as a result of an intentional or accidental nuclear war, we must remember that the explosive power of just one thermonuclear charge is several times greater than the total of all explosions made in the course of all the wars conducted by humanity.

We still remember the tragedy of Hiroshima. But if we take the total capacity of the nuclear arsenals stockpiled in the world, it will be equal to a million bombs of the kind dropped on Hiroshima.

Radioactivity from nuclear weapons is just as devastating for human life as is their destructive power. Even an hour after a one-megaton nuclear explosion, the radioactivity at the place of explosion is equal to the radioactivity of 500 million kilograms of radium. It is scores of millions of times greater than that which is included in the powerful gamma installations used in medicine for the treatment of malignant tumors. Unfortunately, however, few people in the world realize the dire consequences a nuclear war would have for humanity, for each one of us, for our close ones. Figuratively speaking, humanity is now sitting on a powder keg which holds 10 tons of TNT for each one of us. Around this powder keg, certain people are waving the torch of a "nuclear policy strategy" which at any second, even accidentally, may cause an explosion that would be a world catastrophe.

Little by little, with the joys and sorrows of daily life, with the vital problems facing them, people began to forget the horrors of Hiroshima and Nagasaki. Moreover, some of the military, public functionaries and even scientists are trying to downgrade the danger of the nuclear arms race, to minimize the possible consequences of a nuclear war. Statements appear that a nuclear war can be won; that a limited nuclear war can be waged; that humanity and the biosphere will persist, even in conditions of total nuclear catastrophe.

This is an illusion which must be dispelled. Hiroshima and Nagasaki are a reality, are a historical fact which shows that science and technology have fostered power that can lead to global annihilation of every living thing. The bell that tolls in Hiroshima reminds people of the danger. It calls on them to be vigilant. It calls on them to do their utmost to prevent the tragedy of a nuclear war.

Outstanding scientists and theoreticians, physicians who best of all realize the threat to humanity of the nuclear arms race, understood the need to work for nuclear disarmament, for a ban on nuclear weapons. Twenty-five years ago the Pugwash movement of scientists was organized at the appeal of Einstein, Russell, Joliot-Curie and others. Pugwash urged an assessment of the danger that loomed as a result of the creation of weapons of mass destruction. The appeal, which not only pointed to the danger of a nuclear war but was also an appeal to wisdom and to the solution of problems between states by peaceful means.

The Pugwash movement played a certain role in the effort to limit nuclear weapons. Unfortunately, however, its ideas have not reached the wide masses of society, have not penetrated the hearts and minds of everyone on our planet.

We physicians, as the professional group that is the most influential in our struggle for the life and health of all the people in the world, as the most knowledgeable about the tragic consequences of a nuclear war, can make a bigger contribution to the cause of preventing it. Our patients trust us; they entrust their health and their lives to us. And in keeping with our professional honor, in keeping with the oath of Hippocrates, we have no right to hide from them the danger to their lives which now threatens us all.

Today, physicians working all over the world must regard the struggle against the danger of a nuclear war to be not only the duty of an honest, humanitarian person, but also a professional duty. This must be understood by the World Health Organization. We are dealing not with political problems but with the preservation of the health and lives of all people.

Only a year has passed since the idea was born of setting up the international movement of physicians for nuclear disarmament. It is with a feeling of great satisfaction, with a feeling of professional pride for the high moral and humanitarian qualities of physicians in various countries of the world, that we can say this idea has won wide approval.

The reply to our American colleagues, supporting the organization of the movement of physicians for nuclear disarmament, was signed by 90 outstanding Soviet medical scientists and doctors, representing 35 medical societies.

Proof that the idea of the movement is widely supported is seen in the delegates taking part in the work of the conference. Not only are there American and Soviet representatives but also physicians from Japan, Britain, France, Sweden, Canada, Norway, Israel, West Germany and the Netherlands, as well as an observer from the World Health Organization.

We have started to disseminate our ideas widely among the people of our various countries. A television broadcast, in which we spoke of the main principles of the movement of physicians for limiting and banning nuclear weapons, and in which we told of the possible tragic consequences of a nuclear war, was seen by more than 150 million citizens of the Soviet Union. The most diversified political forces support this international movement.

Raising our firm voice of protest against the nuclear arms race and for the prevention of a nuclear war, we must at the same time find the most effective path for our cause. One of our main principles must be the desire to explain to the peoples of the world and to the governments possessing nuclear weapons, on the basis of our knowledge and precise research data, the danger to life on Earth that stems from nuclear weapons and the unleashing of a nuclear war. Moreover, we must discuss not only the immediate consequences of a nuclear explosion, but also all the global problems resulting from the radioactive contamination of the stratosphere—the disruption of the ozone layer of the Earth, the changes of the climate, ecology, and more. It must be emphasized that no country or people will remain unaffected by a nuclear catastrophe.

We must explain to the peoples and the governments that under conditions of nuclear war, medicine will be unable to provide aid to the hundreds of thousands of wounded, burnt and sick, because of the deaths of doctors and the destruction of the transport system, drugs, blood supplies, hospitals and laboratories. Epidemic outbreaks will reach far beyond the affected centers. An important principle of our movement, which the peoples of various countries understand, is the assertion of the fact that already today, the nuclear arms race is very costly for humanity. First and foremost, this is because of the serious psychological harm stemming from the fear experienced by people all over the world as a result of the threat of a nuclear war.

Tremendous sums are being spent on the technology of a nuclear war at a time when millions of people go hungry; when they suffer from various diseases; when illiteracy is still widespread. Huge expenditures on manpower and material resources render difficult the solution of numerous world problems—health, energy, education, economics and more.

In the developing countries today, 100 million children are in danger of dying because of malnutrition and vitamin shortage, and 30 percent of the children have no possibility of going to school. Yet military expenditures worldwide are 20 to 25 times bigger than the total aid provided annually to the developing countries by the developed states. For example, in the past ten years the World Health Organization spent about $83 million on smallpox eradication—less than the cost of one modern strategic bomber. According to some World Health Organization calculations, $450 million are needed to eradicate malaria, a disease which affects over one billion people in 66 countries of the world. This is less than half of what is spent in the world on arms every day and only a third of the cost of a nuclear submarine of the Trident type.

Our principles are the principles of preserving life on Earth, of providing happiness for all, for our children and grandchildren.

People of various political outlooks, nationalities and religions must urge governments to concentrate their attention not on what steps to take to attain victory in a nuclear war, but on what must be done so that the flames of such a war will never burn on our planet.

We face many difficulties, and the path to the achievement of our goal will be thorny and full of impediments. But we have no alternative: when we hear the call that "humanity is in danger" the physicians must rise to the challenge. And we are calling on all the physicians of the world, on all medical workers, to merge their efforts for the salvation of life on Earth.

6 A clear and present danger —West

HERBERT SCOVILLE, JR.

What is the true "present danger" today? In my view, every day, by both actions and words, we are moving closer and closer to a situation where a nuclear war might actually occur. The horrors of a nuclear war are coming closer to reality. That, not the Red Menace, is the real present danger.

Since 1945, nobody has exploded a nuclear weapon in a conflict. There has been built up over the last 35 years a firebreak between conventional and nuclear weapons—a widespread understanding that no matter what the nature of a conflict, escalation to nuclear weapons should not be allowed to occur. The only time nuclear weapons came close to being used was at the time of the Cuban Missile Crisis in 1962; and frankly, I don't think we were so close to it even then. The United States has gone through two major wars, in Korea and in Vietnam, and there was never any serious contemplation of using nuclear weapons in those wars, even though we essentially lost one, and I guess we can say we tied the other.

We have had this firebreak, and strengthened it over time. But, unfortunately, today we are on a slippery slope toward nuclear conflict. The probability that it will actually happen, that somebody will use one of these nuclear weapons in a conflict, or perhaps even by accident, is growing.

An increasing number of people are saying that we can fight a nuclear war, that we can survive, and that we can even win it. The view that such people should be put in an insane asylum is very sound; but that is not going to happen because these people are the leaders of our country and—in some cases—the leaders of the

Soviet Union as well. It is a little more difficult to know the real thinking in the Soviet Union; therefore if I speak more about the United States it is because it is easier to document what goes on here. Nevertheless, we must realize that these dangers are by no means entirely of our own making. The Soviets are not blameless.

Last year, after our frustrations with being unable to do anything about the Soviet invasion of Afghanistan, we heard semiofficial rumors that if the Soviets moved beyond Afghanistan, we would have to consider using nuclear weapons to defend our vital interests in the Middle East and Persian Gulf areas. It's not clear what using these nuclear weapons would accomplish to maintain our oil supplies, but it was thought by political leaders that flexing nuclear muscles might somehow solve an insoluble problem.

The platform of the Republican Party stated that we have to be prepared to fight a nuclear war and implied that we could win it. Then the Democratic President promulgated Presidential Directive 59, in which he specifically directed the U.S. government to prepare to fight a limited nuclear war. This was rather strange, coming from President Carter and his Secretary of Defense Harold Brown, since over the past four years on occasion after occasion, they had both said they know absolutely no way to keep a small nuclear conflict from escalating into all-out nuclear war.

When I was in the government, back in the 1960s, I used to participate from time to time in so-called "war games" that the Joint Chiefs of Staff held. On two occasions we dealt with the European situation. These war games started with the Red Team carrying out some kind of aggressive act in Germany. Then the Blue Team wasn't doing very well, and so it said, "We've just got to show the Reds we are serious." So it dropped one nuclear weapon on a Red Team tank battalion and wiped it out. The Blues thought that was going to convince the Reds to back off and that there would be no further aggression. Unfortunately, the Red Team said "We're not going to be bluffed by this sort of thing; we're going to show them that we really mean business." And they dropped five atomic bombs on five of the Blues' airfields, knocking out the planes. The action then zigzagged back and forth, escalating at each step. The net result of the war game was not only that there was no Europe left; there was also no Soviet Union, no United States, at least as modern societies.

There is just no way one can count on keeping a limited nuclear war limited. That fundamental point must be recognized. Jumping that firebreak between conventional and nuclear weapons is a sure prescription for disaster. We don't want any kind of war, but one of the things that is absolutely certain is that we cannot rely on nuclear weapons to win a conflict.

It's not only the words that are currently disturbing; but it is the actions that are probably even worse. In some cases it's not always easy to see which is the chicken and which is the egg. Far too often policies are devised after the fact, in order to justify some specific weapons program. This might well be the reason we have Presidential Directive 59, which really seems to be an *ex post facto* justification for some of the weapons programs the Carter administration had been backing.

First and foremost is the MX. This is a missile system that is planned to be the follow-on to our existing Minuteman ICBMs, which are supposedly becoming vulnerable in this decade. The MX missile has had much publicity—probably not really because of its true failures and dangers. Rather, it is the Rube Goldberg plans for its deployment that have attracted tremendous public attention. In addition, it has developed strong opposition in Nevada and Utah where it is proposed to be deployed.

The goal of trying to get an invulnerable ICBM force is a worthy one. Personally, I don't think that there is any reasonable possibility that the Soviets could launch an attack against our Minuteman missiles and have any confidence that they could knock them all out. Further, they must keep in mind that even if they were completely successful, we could still retaliate with more than 5,000 warheads, each of which has an explosive force at least three times greater than that of the bomb that destroyed Hiroshima. That they would ever launch such an attack, and count on getting away with it, is to me an unreal notion. Yet we still don't like to see our ICBMs even theoretically vulnerable, because vulnerable systems become attractive targets, especially in times of crisis. If a significant part of a nation's force is vulnerable, then there are enemy incentives to launch a first strike at the missiles before they can be fired and while they are still in their silos. It is this temptation that we would like to avoid. It is why neither side should have a capability to destroy the ICBMs of the other.

To reduce the MX system's vulnerability we propose to play the "shell game" with the Soviet Union. Initially each new missile was to be moved on a "race track" among 23 shelters, the idea being that the Soviets would never know which shelter hid the missile. For the 200 MX missiles there were to be a total of 4,600 shelters. More recently, the race track plan was changed to linear arrays, but the principle is the same.

The principle behind any multiple launch point scheme, however, is basically flawed for two reasons:

- The game of deception is the wrong game to play with the Russians. We would never know whether something we did had given away the location of the missile, or whether the Soviets knew where the missile was. In short, we would never know if we had accomplished what we set out to do.
- The game won't provide survivability unless we can be sure of a continuing arms control agreement, limiting the Soviets to a small enough number of warheads to prevent them from being able to threaten all 4,600 of the shelters. The Soviets would have been allowed 6,000 ICBM warheads under the SALT II Treaty, probably not enough to knock out all 4,300 shelters. But we don't have a SALT II Treaty, and it doesn't look as if we're about to get one. Even if we had, it would have run out in 1986, about four years before the MX system would be fully operational.

Thus, the MX missile, with any type of multiple launch point basing, won't ensure survivability. And without a guaranteed limit on Soviet warheads we will not get the invulnerability that we are seeking.

That alone would be reason enough to oppose spending billions on the MX. (The conservative estimate is $60 billion; $100 billion is more likely.) But the most important consideration, which gets very little attention, is that the MX missile system itself is very dangerous because it increases the risk of nuclear war. The 200 missiles with 2,000 accurate warheads are designed specifically to have the capability to threaten the entire Soviet ICBM force, and ICBMs are 75 percent of their strategic deterrent.

The United States and the Soviet Union must both be confident that they have a survivable deterrent; otherwise we have an unstable situation in which there are pressures to launch a first strike. Yet the MX can only be considered by the Soviet Union as a first strike weapon. It won't destroy missiles in a second strike because there won't be any missiles in the Soviet silos when our warheads arrive: They will have been launched in a first strike, or when the Soviets get warning that we are retaliating.

Since the MX will threaten a large part of the Soviet deterrent, they must do something to reduce the vulnerability. But every way in which they are likely to respond increases the danger that everyone in the United States will be involved in a nuclear catastrophe.

The first and easiest thing the Soviets can do is to put their weapons on a hair-trigger alert, known as "launch-on-warning." When their warning systems tell them that we have launched an attack, in the 15 minutes while the U.S. warheads are on their way, the Soviets can launch their missiles toward their targets. When the U.S. warhead arrives, it hits an empty silo. This is a very cheap and foolproof method of making sure that your missiles are not destroyed. But it is terribly dangerous for the United States.

What it means is that we are depending on computers and radars and other technologies to tell us when we should start a nuclear war. Frankly I don't want my future and the future of the world dependent on whether a computer malfunctions or not. We've seen three cases in the last year in the United States where we had actual nuclear alerts due to malfunctions of our strategic weapons computer system. Fortunately, we are not on a hair-trigger alert policy. But with the MX deployment we are pushing the Soviet Union down that road.

Even if the Soviets don't go to launch-on-warning, we are almost ensuring that in a time of crisis they will launch their missiles in a preemptive strike. Knowing that 75 percent of their deterrent is threatened, they will worry that we will launch first. For example, if a conventional war were to start in Europe, Moscow would be under pressure to launch a preemptive nuclear strike at our ICBMs to forestall any U.S. first strike. The MX increases by a very substantial margin the chances that we will all be involved in a nuclear war, either by accident or through preemption.

But our programs that make nuclear war more likely are not limited to well-known strategic systems such as the MX. We're doing the same thing with weapons in Europe. The probability of a conflict breaking out in Europe is far greater than that of one occurring as a result of direct confrontation between the Soviet Union and the United States. Yet, a year ago, the United States exerted strong pressure on its NATO allies to accept the deployment on land of strategic missiles—the 576 cruise and the long-range Pershing ballistic missiles—aimed at strategic targets in the Soviet Union.

For a long time NATO has had a strategic deterrent force of invulnerable submarine missiles. Since these cannot be located and thus targeted, they do not provide incentives for a Soviet first strike. Thus there is no need to have additional weapons in Europe, even though the Soviets are replacing their old medium-range missiles with new SS-20s. After all, they've had about 700 old missiles aimed at Europe since the mid-1950s; there is nothing new about such a threat.

Now, however, we are basing new weapons on land, where they are far more apt to provoke a Soviet attack. In any conflict in Europe, the Soviet military leaders may decide that they cannot afford to have this Sword of Damocles hanging over their heads and will try to wipe out these missiles before they are fired at the Soviet Union. We will have increased the likelihood that nuclear weapons will be used in Europe, and that a large-scale nuclear war will break out. This example demonstrates that many of the weapons we are proposing to deploy do not buy security. What they do is make it much more likely that we will be involved in an actual nuclear conflict.

The possibility of a nuclear war is not limited to a direct U.S.-Soviet nuclear confrontation. Indeed, the leaders of both countries probably realize the extreme dangers of such a confrontation. But a nuclear war can start elsewhere. The chances are growing that other countries will have nuclear weapons, or that some weapons will fall into irresponsible hands.

A basic and long-standing U.S. policy has been to try to halt the spread of nuclear weapons. President Carter recognized this as a very important security objective of the United States. Unfortunately, although he started his Administration with a major effort to curb the proliferation of nuclear weapons, he later compromised that particular goal for other objectives.

The Afghanistan invasion, for example, was allowed to override the worries about Pakistan building nuclear weapons. Yet that is something that really should frighten us. Pakistan is the most dangerous nuclear weapons "hot spot" in the world today. It seems certain that the Pakistani government has a secret program to produce nuclear explosives. It is building a separation plant to obtain highly enriched uranium-235, which could be used for weapons. It claims that this uranium is for peaceful purposes, but there's no justification for peaceful nuclear explosives in Pakistan. Furthermore, there are reports that the financing for this secret program was obtained from Libyan leader Qaddafi. In

view of such real dangers, we have to treat our nuclear non-proliferation goals more seriously. We cannot abandon them for every other issue that arises.

What are the United States and the Soviet Union doing to slow proliferation? We're never going to halt the spread of nuclear weapons if the two nuclear giants don't start acting with some restraint themselves. Neither country has acted with any restraint despite their agreement to negotiate to limit nuclear weapons since the Non-Proliferation Treaty was signed in 1968. In 1980 President Carter overruled the Nuclear Regulatory Commission and stopped a veto by the Congress with only a few votes, when he sought to sell additional uranium to India. Yet India refuses to guarantee that it will not build nuclear weapons. It has already exploded a nuclear device which it claims is for peaceful purposes. But there is no difference between a peaceful explosive and a weapon. That explosion was the trigger that set off Pakistan's weapons program. Again, our non-proliferation goal was sacrificed in order to try to keep India on our side in the confrontation with the Soviet Union over Afghanistan.

Furthermore, in September 1980, without any publicity, President Carter agreed to an expansion of our production of fissionable material for weapons. We have hundreds of thousands of kilograms of fissionable material in the weapons programs already. We had more than we knew how to use back in 1969, even though we were then contemplating building ABM systems and procuring MIRVs. Yet we now insist that we have to expand the production programs. How does this look to India and other countries which are trying to decide whether they will have nuclear weapons? We say to them: "You can't even have ten kilograms, and you must have all of your plants safeguarded with international inspectors because you might make a bomb with your ten kilograms." Then we turn around and say "Two hundred thousand kilograms is not enough for us."

We have also failed miserably in reaching an agreement to stop all nuclear testing, despite the fact that in the past we and the Soviet Union have between us conducted more than a thousand nuclear tests. We also threaten to use nuclear weapons in a conventional conflict. We will never persuade other countries not to acquire nuclear weapons as long as we don't exercise more restraint ourselves. Restraint may not solve the difficult proliferation problem, but it is absolutely certain that without restraint it will never be solved.

My final point has to do with arms control. Our highest priority should be to put some controls and limits on nuclear weapons so we don't go madly on, getting more weapons incorporating recent advances in technology. These advances are actually making nuclear weapons more usable, and increasing the danger that nuclear war will occur. The SALT II Treaty was not ratified in the Senate, but it wasn't just Afghanistan that killed it. The ratification debate was going very well until August 1980, but then the terrible mishandling of the Soviet-brigade-in-Cuba episode started the downhill process.

The Carter administration, in an attempt to swing votes for the SALT ratification, decided mistakenly to support more and more new weapons programs, the MX being a good example. The reason Carter decided to go ahead with the MX program was in order to buy off the opposition to SALT II. Of course it didn't work. Carter approved the MX program, and in the meantime those who were opposed to SALT sought another bribe. They demanded a 5 or 10 percent increase in the defense budget. The Carter administration mishandled SALT, and SALT II is now dead.

The Reagan administration has talked about a SALT III treaty, but no steps have been taken to follow through. The changes that Reagan advisors have been suggesting are ones that will not be acceptable to the Soviets. Their idea that the Soviets must cut back on their programs while the United States goes ahead with more weapons will be a hard one for our negotiator to sell to his Soviet counterpart. The SALT II Treaty wasn't the greatest in the world, but it formulated some important first steps. It would have been a very useful framework for the future, and it is a great setback to our security that it has gone down to defeat.

Worse still, every other area of arms control is in trouble. Test Ban Treaty negotiations have already resolved most of the technical problems; we could have a treaty tomorrow, given a political will to stop testing. But there are those who push for more nuclear tests, despite the hundreds that we have already conducted. The government is not satisfied that our security would not only be improved if nobody was testing, but also our non-proliferation goals would be enhanced. The Comprehensive Test Ban Treaty is therefore in very serious trouble, and I doubt very much that progress will be made on it in the next four years.

All of the arms control measures are in disrepute, but those who recognize the dangers of a continued nuclear arms race must not give up. Progress need not come only through agreements. Somehow, the United States and, reciprocally, the Soviet Union must start taking national actions, initially in little steps, toward arms limitations. I do not favor unilateral nuclear disarmament; that doesn't work. But we have to start with each country's stopping the procurement of these new weapons that can only increase the chances that a nuclear conflict will actually occur.

7 Instruments of war

K. TSIPIS and JOHN ISAACS

The nuclear strategic arsenals of the United States and the Soviet Union consist of thousands of warheads of varying explosive yields designed to be carried to their targets by a variety of delivery vehicles. Partly by design and partly by circumstance and political expediency, the U.S. strategic arsenal consists of a triad of delivery systems: intercontinental bombers, land-based missiles with intercontinental ranges, and submarine-based missiles capable of reaching any target in the Soviet Union.

U.S. delivery systems

Each nuclear delivery system has its own peculiar advantages and disadvantages, strengths and vulnerabilities. The bombers, for example, can carry about 10,000 pounds of bombs and are recallable even several hours after they have been dispatched to their targets. In the event of a surprise attack, the survivability of the bombers will depend upon how quickly they receive a warning of the launching of the hostile missiles. If these missiles are launched by submarines off the coasts of the United States, for example, the bombers will have about 15 minutes to escape attack. Therefore, it is certain that even if warned immediately, a number of bombers could be destroyed on the ground.

Another vulnerability of the bomber is anti-aircraft defenses. If the bomber has to be over a target in order to attack, it will be subjected to anti-aircraft fire from the ground or from fighter aircraft. Under the worst possible circumstances ever encountered in the past, the attrition rate for U.S. bombers was a few percent (2 to 5 percent) per sortie. In order to minimize this threat to its bombers, the United States uses stand-off missiles that can attack a target from a distance of about 50 miles away, enabling the bomber to avoid the thick defenses around a valuable target. In the future, U.S. bombers will carry long-range cruise missiles that can fly independently for hundreds or thousands of miles and find their target with excellent accuracy. This will make it totally unnecessary for the bombers to overfly enemy territory at all, thus further limiting their vulnerability to anti-aircraft defenses.

U.S. land-based intercontinental ballistic missiles carry about a quarter of the payload of a bomber. But these missiles can fly to their targets in about 30 minutes and are securely kept in concrete silos, which are buried in the ground and designed to protect the missile from all effects of a nuclear explosion that occurs beyond a few hundred feet of the silo. Controlled from reinforced concrete bunkers, these missiles are considered the swiftest means of retaliating against a country that has launched a nuclear attack against the United States. Because of their tight command and control and good accuracy they are thought of as suitable for a "flexible" or a "limited" response to a nuclear attack against the United States. Therefore, their degree of utility in the U.S. triad depends to some extent on the kind of strategic scenario or doctrine one assumes the United States must employ in case of a nuclear war. There are no significant defenses against them.

The third delivery vehicle of strategic nuclear weapons are missiles that are carried by submarines specially fitted with tubes in which the missiles are carried (16 or 24 per submarine), and from where they are launched. A submarine is totally invisible and therefore untargetable once submerged. Therefore, the submarine-carried missiles are by far the most secure part of the U.S. strategic arsenal because they enjoy excellent pre-launch survivability. These weapons will survive any surprise attack against the bombers and the land-based ICBMs, and thus they are the part of the strategic triad that guarantees that the United States can deter any nuclear attack against it. Three-quarters of all U.S.

missile-carrying submarines are always out at sea, and since each carries 16 missiles with about 10 nuclear warheads each, it is apparent that an enormous number of U.S. nuclear warheads are at all times safe and ready to attack an opponent.

The disadvantage of the submarine-based missile system is that it is more difficult to communicate with submarines than with bombers or the bunkers that control the land-based ICBMs. Electrical signals, such as radio waves, do not easily penetrate seawater, and therefore once a submarine is submerged it must either trail an antenna or carry some other means of communicating with the national command post on land. Communications are possible but not easy and they are somewhat laborious, more like the Morse code telegraph than like the telephone or the radio. So although the submarine-based missiles are ideally suited for a retaliatory second strike since they are safe from a surprise attack, they cannot contribute effectively to a counterforce attack against Soviet ICBM silos or other time-urgent targets since such an operation requires a highly coordinated attack.

The first submarine capable of carrying ballistic missiles was the *George Washington*, commissioned on December 30, 1959. It and other submarines of its class displace 6,700 tons submerged and have an approximate speed of 30 knots while traveling underwater. It is navigated with the help of three inertial navigation systems that permit the submarine to remain submerged for long periods of time and still know where it is. Each of these submarines carried 16 Polaris missiles, each armed with 3 warheads with an explosive yield of 220 kilotons (thousands of tons of TNT equivalent). The circular error probable (or CEP) of these warheads was half of a mile. Circular error probable is a measure of the accuracy of the missile. If one performs a large number of tests with the same type of missile aimed at the same target and records the points of impact of each missile, it is possible to define the "mean point of impact." Thus, CEP is the radius of a circle centered at the mean point of impact that contains 50 percent of the impact points. However, the mean point of impact is not necessarily the target point at which all the missiles were aimed.

The United States has a larger class of missile-carrying submarines that displace 8,250 tons submerged and carry 16 Poseidon missiles. Each of these has 10 to 14 warheads with an explosive yield of 100 kilotons and a CEP of a quarter of a mile. There is a variety of Poseidon missiles (C-3, C-4) that have different ranges. Depending on the number of warheads it carries a Poseidon missile has a range of about 4,000 kilometers.

The United States has just launched the first of a new type of submarine, the Trident, that displaces 18,700 tons submerged and has 24 missiles, each armed with 8 warheads with an explosive yield of about 100 kilotons and a CEP somewhat better than a quarter of a mile. Since the first Trident submarine has just begun sea trials there is little information available about its performance.

The land-based ICBM force of the United States consists of 54 Titan and 1,000 Minuteman missiles kept in concrete silos that are able to withstand pressures exceeding 1,000 psi. The Titan missile weighs about 330,000 pounds, has a useful payload of about 8,000 pounds, a range of 6,300 nautical miles and carries a warhead of 5 megatons (millions of tons of TNT equivalent). The Minuteman missiles have gone through repeated upgradings and alterations since they were first deployed in 1960. Minuteman III is a 70,000-pound missile with solid fuel that has an approximate useful payload of 2,500 pounds and a range of 6,000 nautical miles. The 450 Minuteman II missiles carry a single 1-megaton warhead and have a CEP of 0.2 nautical mile. Minuteman III missiles have been undergoing further improvement. Some of them carry 3, Mark-12 warheads, each with explosive power of 170 kilotons and a CEP of 0.12 nautical mile. Some, however, carry an improved Mark-12A warhead with twice the explosive yield and the same precision. All Minuteman missiles have a range that exceeds 6,000 nautical miles. A new experimental missile is now under development in the United States—the MX. In its present design, the MX is expected to weigh about 200,000 pounds, carry 10 warheads, each with an explosive yield of about 300 kilotons, and a CEP of 300 to 400 feet. It is entirely too early to tell whether the missile in its final form will be similar to its current design.

The strategic bomber force of the United States consists of FB-111A intermediate-range aircraft based in Europe, and the B-52 heavy bomber capable of long ranges. The FB-111A has a crew of two, can fly at over 60,000 feet at 2.5 times the speed of sound and has a range of 3,300 nautical miles. It can carry up to 35,000 pounds of munitions, that is, either 6 nuclear bombs or 6 short-range attack missiles (SHRAM) that allow the plane to avoid approaching its intended target. Each short-range attack missile carries a warhead with a 200-kiloton explosive yield and each gravity bomb has about 1-megaton yield. There are about 100 FB-111A's stationed in Europe well within range of the Soviet heartland.

The B-52 bomber was first introduced into the U.S. strategic arsenal in 1954 and has undergone at least seven stages of improvement. It has a crew of six, can fly over at 50,000 feet at about the speed of sound, and has a range of about 8,600 nautical miles. The latest version of the plane (G and H series) carries 20 short-range attack missiles, or an equal number of the long-range cruise missiles now under development. (Alternatively, they could carry 4 to 6 gravity bombs in the megaton explosive yield range.) The B-52 bombers have been modified in order to be able to fly at almost tree-top altitudes in order to avoid enemy radar and anti-aircraft missiles. They have also been equipped with fast start-up generators so that, in principle at least, the entire alert position of the bomber force can be airborne within 15 minutes.

A similar but not identical process of development and evolutionary improvement characterizes the build-up of the Soviet strategic forces.

Soviet delivery systems

The first of the "Hotel II" class of USSR ballistic missile submarines was launched in 1960. It displaced 4,100 tons submerged and carried 3 single-warhead missiles of one megaton each with a range of 700 nautical miles, a throwweight of 2,000 pounds and a CEP measured in miles. In 1968 the USSR launched the first "Yankee" class submarine that approximated the U.S. Polaris. Each Yankee submarine carried 16 liquid-fuel missiles, displaced about 9,000 tons submerged and had a speed of 25 knots. Each missile carries one warhead with an explosive yield of about 1 megaton and has a range of 1,300 nautical miles and a CEP of about 0.5 nautical mile. Because of the relatively short range of their missiles, both the Hotel and Yankee class subs would have to penetrate into the central Atlantic (south of Greenland-Iceland-U.K. line) in order to launch attacks against the United States. In view of the effective anti-sub warfare operations of the U.S. Navy in the northern Atlantic and in the sealanes between Greenland, Iceland and United Kingdom, it is expected that these classes of Soviet submarines would suffer considerable attrition in time of war.

Cognizant of this handicap, the USSR developed a new class of submarine-launched ballistic missiles and submarines: the "Delta" class submarine and the SS-N-8-II ballistic missile.

The first few Delta submarines, displacing about 9,000 tons submerged, carried 12 SSN-8 liquid-fueled ballistic missiles with ranges up to 4,000 nautical miles and a payload of about 2,000 pounds. Each missile has a single one-megaton yield warhead. Eleven Delta I submarines were deployed in 1975 to 1976. The Delta II class submarines are equipped with 16 SS-N-8 missiles that have a CEP of about 0.5 nautical miles. Twenty Delta II submarines were deployed between 1976 and 1977.

Although the Soviet Union has 85 ballistic missile-carrying submarines, they carry only about 20 percent of the strategic weapons of the Soviet Union. At no time have more than 15 percent of these submarines been at sea simultaneously. Therefore, each day only about 2 to 3 percent of the Soviet strategic weapons are safely deployed at sea. The rest of the submarines are docked at pens that would be vulnerable to a surprise attack by the United States. In comparison, while the United States has only 41 ballistic missile submarines, they carry about 50 percent of the U.S. strategic weapons. Of those 41, at least 30 are on station (submerged somewhere in the ocean) at all times. Therefore, about 35 percent of all U.S. strategic weapons are safe from a surprise attack, since once submerged a U.S. missile submarine has never been found or trailed by a Soviet craft.

The Soviet Union has recently launched a new class of ballistic missile submarine, the "Typhoon," which is comparable in size to the U.S. Trident submarines. The Typhoon can carry 24 missiles. They will probably be the SS-N-18. This missile carries 7 warheads of 200-kiloton explosive yield with a reported CEP of 0.3 nautical mile.

The Soviet Union has placed great emphasis on land-based ballistic missiles in its nuclear strategic force. Since 1961 they have developed no less than almost two dozen types of missiles. Of those only a few are deployable.

The Soviet Union has four "design bureaus" that develop missiles in competition with each other. So it is common to witness the development of several similar missiles and the deployment of types that are not as good as others of the same generation. This could be a reflection of internal techno-bureaucratic squabbles on the one hand, and the lack of confidence by the leadership in the ability of a single bureau to develop successfully a new missile. Thus, the continuing redundancy of effort which manifests itself in the four bureaus simultaneously developing new ICBMs. The result is that there is apparently a very large number of engineers and scientists in the Soviet Union that keep working on ever newer models of missiles. As a consequence, the

Soviet ICBM program has both depth and momentum.

Their first truly intercontinental missile, the SS-7, was first deployed in 1962. It had a range of about 6,000 nautical miles, carried a single warhead with a yield of about 3 megatons and had a CEP of about one nautical mile. It is not known whether this missile had an inertial guidance system as all modern ICBMs do. It and all subsequent Soviet ICBMs are liquid-fueled. That means that it takes 24 to 36 hours to prepare a missile for launch, but once ready it can be launched just like the solid fuel Minuteman missiles of the United States. The SS-8 was first deployed in 1963 and is almost identical to the SS-7.

The SS-9 first deployed in 1967 has almost twice the throw-weight of the SS-7/SS-8 family—10 to 15,000 pounds as compared to 8,000 pounds. It is also a liquid-fueled missile and can carry either a 20-megaton warhead or 3 untargetable 3.5 megaton warheads with a CEP of about 1 nautical mile over a range of 6,500 nautical miles.

At about the same time the Soviet Union started deploying still another missile, the SS-11. It has a payload capacity of 2,000 pounds, carries a single warhead with a yield of 1 megaton and has a range of 5,600 nautical miles and a CEP of about 1 nautical mile. In 1973, the SS-11 was tried in tests with 3 untargetable warheads. The SS-13, first deployed in 1965, is also a 2,000 pound missile; it can go only 4,500 nautical miles and carry a 0.75-megaton warhead. Obviously, the bureau that designed it was the least successful of the competing teams. Yet the SS-13 was deployed together with the much more successful SS-11.

A new generation of four classes of missiles (SS-16, SS-17, SS-18, SS-19) began appearing in 1975.

The SS-16 was tested but saw no deployment. It was a solid-fueled missile and could have been a mobile missile.

The SS-17 was first deployed in 1976. It carries 4 warheads, each with a yield of 0.75 megatons, over a range of about 5,000 nautical miles and has a CEP reported at 0.2 nautical miles.

The SS-18 is another giant Soviet missile, similar to the SS-9. It has a payload of 15,000 pounds, a range of 5,500 nautical miles and a CEP about 0.2 nautical miles. There are numerous variants of this missile (see Table 1, p. 86), and most variants carry 8 warheads, each with a yield of 0.75 megatons. (Some sources list the SS-18 warheads as having a yield of 0.6 megatons and those of the SS-17 and SS-19 as 0.4 megatons.)

Obviously, the Soviet Union is experimenting with different configurations of warheads, given the large payload capability that allows them considerable latitude of choice. Without the SALT II constraint that mandates that no more than 10 warheads can be carried by a land-based ICBM, the SS-18 could be configured to carry up to 30 small warheads.

The SS-19 is approximately similar to the SS-17. It can carry 7,000 pounds of useful payload, has a range of 5,000 nautical miles and a CEP reported as 0.17 nautical miles. It carries 6 warheads, each with a yield reported as 0.4, 0.6 or 0.75 megatons.

As a consequence of the great emphasis on ICBMs about 75 percent of Soviet strategic nuclear weapons are carried by these land-based missiles. In comparison only about 20 percent of the U.S. nuclear strategic arsenal is carried by ICBMs.

The Soviet Union has never displayed any serious interest in intercontinental bombers. The two long-range aircraft capable of carrying nuclear weapons over intercontinental distances—the "Bear" and the "Bison"—were first deployed in 1956. The Bear is a propeller-driven aircraft with a cruising speed of about 400 nautical miles per hour and an altitude of 40,000 feet. It can carry about 20,000 pounds of bombs, usually 4, 1-megaton weapons. The Bison is a jet aircraft that can cruise at about 0.8, the speed of sound, and carries the same number of bombs as the Bear. Both of these aircraft have ranges of over 5,000 nautical miles.

In 1974, the Soviet Union deployed a new supersonic bomber, the "Backfire." It has a range of about 3,500 nautical miles and can fly at twice the speed of sound at high altitudes and cruise at

0.6, the speed of sound near the ground. It has a maximum payload of 20,000 pounds and can carry either two air-to-ground missiles under its wings, or 15, 1,000-pound bombs (non-nuclear) in the bomb bay. It is thus a somewhat smaller plane than the U.S. FB-111-A that cannot fly at supersonic speed near the ground. (The FB-111-A can fly at 1.2, the speed of sound at sea level.)

Table 2 attempts to give a comparison of the capabilities of the strategic weapons of the United States and the Soviet Union. Note that many of the figures listed are either projections or, especially in the case of the yield and CEP of Soviet weapons, are not the only ones in the unclassified literature. The figures in this table in general tend to overestimate the performance characteristics of Soviet weapons. Therefore they must be used and quoted with caution.

At any rate, the basic conclusion that one can derive from all these figures is that both the United States and the Soviet Union have very large, diversified and ever improving arsenals of nuclear weapons that can assuredly destroy everything in both countries many times over.

Table 1

Strategic weapon capabilities of the United States and the Soviet Union

System	*Megatons per warhead (y)*	*Number of warheads (n)*	*Total megatons (y•n)*	*EMT*[a] $n \cdot y^{2/3}$	*CEP*[b] *(nmi)*	*TW*[c] *(Klb)*
U.S. DELIVERY SYSTEMS						
Land-based missiles						
Titan II	9.0	1	9.0	4.32	0.7	8.3
Minuteman	1.2	1	1.2	1.13	0.2	1.6
Minuteman III (MK-12)	0.17	3	0.51	0.922	0.2	2.4
Minuteman III (INS-20/MK-12)	0.17	3	0.51	0.922	0.12	2.4
Minuteman III (INS-20/MK-12A)	0.335	3	1.0	1.448	0.12	2.4
Submarine-based missiles						
Polaris A 3	0.22	3	0.66	1.09	0.5	1.1
Poseidon C 3	0.04	9	0.36	1.05	0.25	3.3
Poseidon C 4	0.1	8	0.8	1.73	0.25	2.9
Trident	0.1	8	0.8	1.73	0.25	2.9
Intercontinental bombers						
B-52 (with gravity bombs)	1.	4	4	4	0.1	9.6
B-52 (with air-launched Cruise missiles)	0.2	20	4	6.85	0.08	9.6
B-52 (with short-range attack missiles plus gravity bombs)	0.2+1	12 (8+4)	5.6	6.74 (2.74+4)	0.2/0.1	9.6

SOVIET DELIVERY SYSTEMS

SS-7/SS-8	3.0	1	3.0	2.08	1.0	4.0
SS-9	20.0	1	20.0	7.35	0.5	13.5
SS-9	3.5	3	10.5	6.91	1.0	13.5
SS-11	1.0	1	1.0	1.0	1.0	2.0
SS-13	0.75	1	0.75	0.826	1.0	1.0
SS-18 single	20.0	1	20.0	7.35	0.2	16.0
SS-18 (1977)	0.75	10	7.5	8.26	0.2	16.0
SS-18 (June 18, 1979)	0.75	10	7.5	8.26	0.17	16.0
SS-19 (1977)	0.75	6	4.5	4.96	0.2	8.0
SS-19 (June 18, 1979)	0.75	6	4.5	4.96	0.17	8.0
SS-19/SS-17	15.0	1	15.0	6.07	0.2	7.0
SS-17 (1977)	0.75	4	3.0	3.3	0.2	7.0
SS-17 (June 18, 1979)	0.75	4	3.0	3.3	0.17	7.0
SS-N-6	1.0	1	1.0	1.0	0.5	1.6
SS-N-6	0.5	3	1.5	1.89	0.75	1.6
SS-N-8	1.0	1	1.0	1.0	0.5	1.8
SS-NX-17	0.75	1	0.75	0.826	0.25	2.0
SS-N-18	0.2	7	1.4	2.39	0.3	2.5

Source: U.S. Senate, *Congressional Record,* July 20, 1979, S10078.

[a] Equivalent megatonnage
[b] Circular error probable (nautical miles).
[c] Throwweight (thousands of pounds).

Table 2

Strategic forces of the United States and the Soviet Union

UNITED STATES

Land-based Intercontinental Ballistic Missiles

The United States now has deployed 1,054 ICBMs with a total of 2,154 warheads, including:

450 Minuteman II missiles—each with 1 warhead for a total of 450 warheads;

550 Minuteman III missiles—each with 3 warheads for a total of 1,650 warheads;

54 Titan missiles (although 2 are currently inactive)—each with 1 warhead for a total of 54 warheads.

Future: The United States plans to build and deploy by 1989 200 MX missiles, each with 10 warheads for a total of 2,000 warheads.

Sea-based Ballistic Missiles

The United States as of January 1981 had a total of 36 submarines in the inventory with a total of 576 missiles with a total of 4,816 warheads, including:

5 Polaris submarines—each with 16 A-3 missiles, each missile with 3 warheads, for a total of 240 warheads
(These 5 submarines are slated for deactivation as strategic submarines);

19 Poseidon submarines—each with 16 Poseidon missiles, each missile with about 10 warheads, for a total of about 3,040 warheads;

6 Poseidon submarines—each with 16 Trident C-4 missiles, each missile with about 8 warheads, for a total of about 768 warheds;

6 Other Poseidon submarines are in the process of converting from Poseidon missiles to Trident missiles, which means an additional 768 warheads.

Future: Under construction is a program to build at least 15 Trident submarines—each with 24 missiles, each missile with 8 to 10 warheads.

Strategic Bombers

The United States currently has deployed 412 long-range bombers with a total of about 1,900 nuclear weapons, including:

347 B-52 bombers—each with an average of a little over 5 weapons per plane, for a total of about 1,640;

65 FB-111 bombers—each with an average total of about 4 nuclear weapons, for a total of about 260;

Future: The United States is in the process of equipping 172 B-52 bombers with 20 cruise missiles each. In addition, the U.S. Air Force intends to build a follow-on manned bomber to replace the B-52s, either a version of the B-1 bomber and/or a newer "stealth" bomber.

In sum: In terms of overall strategic systems and weapons, the U.S. strategic forces are comprised of about 1,630 missiles and 412 long-range bombers with a total of about 9,000 nuclear warheads and weapons.

SOVIET UNION

Land-based Intercontinental Ballistic Missiles

The Soviet Union has deployed 1,398 ICBMS, with roughly 5,000 warheads, including:

580 SS-11s—with 1 warhead per missile;
60 SS-13s—with 1 warhead per missile;
About 150 SS-17s—most of which have 4 warheads and some 1 warhead;
308 SS-18s—most with 8 to 10 warheads and some with 1 warhead;
About 300 SS-19s—most with 6 warheads and some with 1 warhead.

Sea-based Missiles

The Soviet Union has about 62 submarines in inventory with a total of 950 missiles and under 2,000 warheads on a variety of submarines.

Strategic Bombers

The Soviet Union currently has about 149 long-range bombers, including:

49 Bison bombers (soon to be phased out);
100 Bear bombers.

In sum: Estimations of the Soviet Union's strategic forces are not as exact as numbers of American forces. Their forces are roughly comprised of about 2,350 missiles and 150 long-range bombers with a total of about 7,000 nuclear warheads and weapons.

III PATHOGENESIS

"Peacetime expenditures on large sectors of the defense establishment are maintained at a wartime level; and the effect is felt in civilian life on a global basis."—J. Carson Mark

We must build more weapons to keep the peace.
Who'll keep the weapons?

8 Nuclear weapons: characteristics and capabilities

J. CARSON MARK

Between March and August of 1945 Japan was subjected to air attacks of unmatched ferocity. There was the great fire raid on Tokyo on March 9, when some 300 B-29s dropped 2,000 tons of bombs. About 16 square miles were burned out and 267,000 buildings were destroyed. A million persons were left homeless, 84,000 died, and 40,000 were injured.

Five weeks later, on April 13 and on April 15, B-29s again bombed Tokyo, dropping around 2,000 tons of bombs, burning out a further 17.5 square miles, but imposing a much smaller number of casualties than did the March 9 raid. Then, on May 23, 520 B-29s dropped 3,650 tons of bombs and, along with another 500 B-29s and 3,250 tons of bombs on May 25, burned out another 22 square miles.

Starting about the middle of June, along with heavy and repeated raids on the main cities of Nagoya, Osaka, Kobe and Yokohama, 58 secondary cities with a total population of 6.5 million were subjected to incendiary attacks, and about 60 square miles of built-up urban areas were destroyed.

In all, between March and the end of the first week of August, 104,000 tons of bombs were dropped on urban areas, along with some 30,000 tons directed at industrial targets. Of this, 43,000 tons were dropped in July alone, and plans were laid to increase the monthly total to 100,000 tons before November 1945. Including those in the major cities, about 8.5 million persons were displaced.

Civilian casualties resulting from these "conventional" attacks numbered about 700,000, of whom some 250,000 were killed.

About 35,000 separate B-29 sorties were conducted, and because of the weakness of Japan's air defenses, only about 500 B-29s were lost. The total tonnage of bombs dropped on Japan was only about 10 percent of that dropped on Germany in the course of the war, but because of differences in the nature of the cities and the strength of air defenses, the population disruption was much greater in Japan.

During the final weeks of this phase of the attack on Japan several medium-sized cities were removed from the list of targets for conventional air attack to provide a reserve of possible targets for the "special" bombs expected to become available in August. On August 6, Hiroshima was destroyed by the first nuclear weapon used in combat, followed by the August 9 attack on Nagasaki. On August 14 the Japanese accepted terms of surrender as stipulated by the United States at Potsdam on July 26.

The physical damage from these explosions, of 12.5 to 20 kilotons, was, of course, enormous; but it was not unprecedented. At Hiroshima the area destroyed was about five square miles; at Nagasaki, only about two. The difference, including the difference in numbers of fatalities, was accounted for partly by the presence of a larger harbor area near the aiming point at Nagasaki, and because the hills and ravines on which Nagasaki was built gave it an irregular pattern of separate populated areas.

The number of civilians who were killed immediately or died within 30 days of the attack was very large: about 70,000 at Hiroshima, and 40,000 at Nagasaki. But again, this was not unprecedented, since the number who died as a result of the March raid on Tokyo was certainly larger, as was the figure for civilian dead as a consequence of the raid on Dresden in February 1945. That number is believed to be greater than 100,000, but it will never really be known because the normal population of Dresden, about 600,000, was almost doubled by the influx of refugees from the East.

Enormous physical damage and frightful numbers of casualties are commonly taken to be distinctive features of nuclear weapons. But they were not outside the range of what could be accomplished by conventional weapons, even those available 35 years ago, and 100 kilotons of bombs a month were showered by U.S. forces on portions of Viet Nam during most of 1972.

The radiation accompanying nuclear explosions was, of course, entirely new and insidious, and this has come to be taken as the main horror associated with nuclear weapons. In Hiroshima and Nagasaki, however, radiation was not a dominant cause of death or acute injury. Of the injured survivors about 70 percent suffered from the blast effects, 65 percent from serious burns, and 30 percent from radiation. (Many, of course, had injuries of more than one type.)

It is unlikely that the Japanese would have terminated the war when they did merely in response to the physical destruction and civilian casualties resulting from these attacks.

The staggering effect of the nuclear weapon resulted from its yield-to-weight ratio and the fact that this level of destruction could be imposed by the explosion of a single package dropped from a single plane. This, along with the fact that a repetition of such attacks could be expected, did, however, provide the leverage needed by those members of the War Council who, since the defeat of Germany in May, had been convinced that Japan should sue for peace.

One of the most important and distinctive features of nuclear weapons is the very large reduction in the cost of imposing a given amount of damage: the cost of the explosive itself, and of the planes, men and effort to deliver it. It suddenly became appallingly easy for anyone possessing nuclear weapons to achieve as much destruction as would previously have required a large fleet of large planes. At the end of World War II, only the United States had access to such weapons; but now many others have, or could have them.

Another difference is that an attack can be accomplished in an instant. Thus, those exposed at the moment of the attack have no chance to take any evasive or protective action. In contrast, during heavy air raids with conventional bombs people could seek shelter even after the start of the bombardment, which usually required several hours to conduct. With a single large nuclear explosion, severe or complete destruction extends outward from the center in a sheer sweep to the radius, where the effects become moderate and the damage sporadic. This could seem particularly advantageous to an adversary trying to destroy an industrial target under conditions similar to those which had applied in World War II. There, the radius of damage from a conventional bomb was smaller than the aiming error, and a number of properly distributed conventional bombs would be needed to ensure destruction of the target. This could be corrected by a large explosion; but the effects of a large explosion are completely indiscriminate. Wherever the incidence of serious injuries is very high there will be few, if any, able to offer assistance. In the particular example of Hiroshima, 270 out of a total of 298 doctors were killed along with 1,645 out of 1,780 nurses. And 42 out of 45 hospital facilities were destroyed or rendered useless.

Less obvious, but also of great importance, is the thermal radiation which accounted for incendiary effects and flash burns to exposed persons. As a result of the enormous amount of energy in a small mass the temperature is initially extremely high—about 10^7 degrees K (K being the absolute scale of temperature in which zero equals -273.1 degrees centigrade). As this material expands, and the energy is shared with an increasing amount of surrounding air, it cools. All the while it emits x-rays in a spectrum appropriate to the temperature, and the rate of energy radiated decreases as the fourth power of the temperature.

At first, while the temperature is very high, these x-rays are absorbed in a few centimeters of air, and the radiation serves merely to engage more material in the heated zone. By the time the mass and volume heated has cooled to about 10,000°K—and by this time the major fraction of the hot material consists of air—the wave-lengths of the thermal radiation are in the range of visible light to which air is transparent. The energy radiated in this phase is transmitted freely through the air without heating it appreciably, until it falls on some solid absorbing surface. There, if the energy density is high enough, (or the surface close enough), it may induce ignition in such flammable materials as wood, upholstery or clothing, or cause serious burns on exposed skin. This process continues until the temperature of the fireball drops to about 2,000°K, by which time the rate of radiation of energy has become so low—in competition with hydrodynamic processes—that it is no longer significant.

Through this phase, about one-third of the total energy of the explosion has been distributed as thermal radiation; and about a ton of air per ton of yield will have been heated to a temperature of 2,000°K or more. About 1 percent of the air heated in this way is converted to nitric oxide. In conventional explosions the temperature of the burned material is only about 3,000°K before it expands at all. No significant fraction of the energy is radiated, nor is any appreciable amount of the surrounding air heated to as high a temperature as 2,000°K. This process accounts for the strong incendiary effects of a nuclear explosion and for the fact that there were about as many deaths and serious injuries from burns as from blast effects.

The atomic weapons dropped on Japan had yields of about 20 kilotons, and the three prompt effects—blast, thermal and nuclear radiation—were each separately hazardous (that is, likely to cause fatalities in an appreciable fraction of persons exposed without protection) out to a radius of about 1.5 kilometers. Though a considerable number of persons inside that radius survived, a comparable number were killed at greater distances. At each city the number of dead was close to the number initially present inside that radius.

In present-day arsenals there are many weapons with yields

considerably larger than 20 kilotons, and also many with smaller yields. For blast effects, the variation with yield of the radius at which a stated overpressure is realized goes closely with the cube root of the yield. For thermal effects, the distance at which a given number of calories per square centimeter is delivered is proportional to the square root of the yield. But, because the thermal pulse from a large explosion has a longer duration than that from a small explosion, the radius of thermal damage does not increase quite so rapidly as the square root. Since the intensity of prompt radiation is attenuated exponentially in air, the distance to which a given intensity is delivered increases only about 0.3 kilometers for each factor of three increase in yield. Taking 1.5 kilometers as the serious hazard radius for all three effects at 20 kilotons, the corresponding radii for one kiloton—such as might be considered for a tactical, or battlefield, weapon—would be: for blast about .55 kilometer; for thermal, about .35 kilometer; for prompt radiation hazard, about .7 kilometer.

For weapons much smaller than 20 kilotons the radiation hazard radius is, then, the largest. For the so-called "neutron bomb," if the blast effects were kept constant, the radiation radius might be larger by .3 or .4 kilometer; or, if the radiation intensity were maintained, the other radii would be smaller by one, or a few, tenths of a kilometer. These radii, of course, are for hazard to exposed personnel, and the radii for destroying a tank or incapacitating a tank crew would be considerably smaller—although the prompt radiation would probably still have the largest range. For comparison, a megaton bomb (such as might be associated with a strategic missile) would have radii for hazards similar to those at 1.5 kilometers from 20 kilotons of: 5.5 kilometers for blast; about 9 for thermal; but only on the order of 2.6 for prompt radiation. It will be seen that the prompt radiation which, though appreciable, was not dominant at 20 kilotons, will be of little or no importance, since very severe blast and thermal hazards extend to much larger distances.

For a megaton the areas covered by severe blast and thermal effects are, respectively, about 15 and 35 times larger than they were for 20 kilotons. If the population density and distribution were similar over these larger areas, the number of fatalities in the event of an attack without warning might be something like 25 times larger—that is, between one and two million. Judging from past experience, the number of injured survivors would be comparable. There are many cities, or metropolitan areas, which could meet these criteria. The great majority of the injured survivors would suffer from burns. The successful treatment of very serious burns requires highly specialized facilities. These are quite limited in the United States, existing primarily in large urban centers.

Obviously, the possibility of a million or so civilian dead from a single bomb on a large city is completely off the scale set by the largest conventional air raids. It is large even compared to the worst recorded natural disasters, such as the 1970 cyclone in East Pakistan (some 500,000 dead), or the 1887 flood on the Hwang Ho with its estimated 900,000 dead. It is more nearly on the scale of the number of battle deaths experienced in World War II by some of the major participants, such as Japan's 1.5 million or Germany's 3 million. The number of deaths would be exceeded by only a few of the recorded plagues and epidemics: the Black Death of the fourteenth century, or the influenza epidemic of 1917 to 1918, with its 20 to 30 million victims. However, it would require only a dozen or so such bombings to come up to these levels; and the number of fatalities from a full-scale nuclear attack on either the United States or the Soviet Union is usually estimated to be of the order of 100 million.

There is some hope that efforts to eliminate the threat of great epidemics may already have succeeded. The threat posed by nuclear war is still with us.

In addition to their effects on people, there are technical implications associated with nuclear weapons. Through progressive improvements in design since the earliest models, the size and weight of nuclear weapons were greatly reduced, while the yield was greatly increased. By the late 1950s it had become possible to package a megaton in a ton. The extremely important consequence was that the weight was low enough so that it could be carried over intercontinental distances in missiles of manageable proportions, while the yield was high enough so that against normal structures its damage radius was larger than the miss-distance. The yield-to-weight ratio of nuclear weapons thus enabled the initial deployment of a strategic ballistic missile force.

Very great improvements in missile guidance over the past couple of decades have made it possible to plan to destroy quite hard (blast-resistant) targets with yields much smaller than a megaton. Continued progress in reducing the size and increasing the yield-to-weight ratio of weapons in the few tens or hundreds of kilotons has made it possible to put several warheads in a single missile and have these directed to separate targets. The stockpiles of the United States and the Soviet Union now contain several thousand such weapons, and, as well as being frightfully capable against urban and industrial targets, it is believed that the accuracy is now such that they can destroy land-based enemy missiles in their hardened silos.

By combining this scenario with the theorem that in the modern world lasting happiness can only be pursued in the lee of a redundant and overwhelming missile force, it is now said that we have an urgent need for an improved and mobile missile force—the MX. One thing is certain; left to itself, this iterative process will continue.

Besides directly attacking other weapons in their emplacements—as well as ships, planes, military installations, dams, essential industrial facilities, people and all their civilian works—nuclear weapons can be used for other disruptive effects. For example, radiations from nuclear explosions at altitudes above a few tens of miles create clouds of highly ionized air or impose changes hydrodynamically in the ionosphere. These may interrupt long-range communication links, or even the capabilities of radar observation. Effects of this kind were observed following the high altitude tests by the United States over Johnston Island. Additionally, because of the low air density at high altitudes the range of all the radiations from a nuclear explosion becomes very large. At considerable distances these radiations, impinging on a satellite or nuclear warhead, or its carrier, could make their outer surface (or some inner layer of fissile material) hot enough to damage or destroy the target.

Another point of major current concern is that megaton-like bursts at altitudes above about 100 kilometers would create an intense electromagnetic pulse which could result in electric fields, over a large part—or even all—of the continental United States, in excess of 25,000 volts per meter. In antennas—whether long communication lines or power lines, or even the cables associated with electronic equipment on planes or missiles—these voltages could induce currents which would destroy the transistorized circuitry on which modern versions of such systems rely for their functioning, unless they should be adequately shielded at all points against such effects.

Of particular concern are the communication links for the command and control of our missile-launching complexes as well as the guidance systems for the missiles themselves. Though of less strategic importance, the civilian alarm and confusion which would follow a widespread and extended breakdown of the U.S. power grid, with the cessation of all electric services lacking a separate power supply—communication, lighting, ventilation, pumping (of water as well as gasoline), much transportation, to say nothing of elevators—would be spectacular.

Even in peacetime, nuclear weapons have had a profound effect. For 30 years now, war—at least as between the superpowers—has been pictured as taking the form of a nuclear blitzkrieg in which a large number of weapons would be exchanged in a short time and enormous damage inflicted on opposing weapons systems or civilian targets; or, more probably, on both. For such an event only weapons existing at the start would be of any account. Consequently, it has been argued that it was necessary to stockpile in peacetime all the armament that might be needed in the event of war. This is in contrast with previous wars, when full-scale armament production was undertaken only after the start of hostilities. In addition, it has been shown that technology can be relied on, not merely to develop improvements in weapons systems but to uncover possible new vulnerabilities. It has thus become necessary to add to or replace parts of the stockpile on a continuing basis with more advanced and more expensive models. As a response to the existence of nuclear weapons, then, peacetime expenditures on large sectors of the defense establishment are maintained at a wartime level. This effect is felt in civilian life on a global basis.

Another matter of deep concern is the residual radiation associated with fission. In the act of fission the uranium or plutonium nucleus releases two or three neutrons and two major fragments: a lighter one with a mass of about 100, or a little less, and a heavier one with a mass of 130, or a little more. While the total mass of the parts is essentially constant, there is considerable variation in the masses of the two pieces. The result of a large number of fissions is thus a mixture of isotopes of a great variety of lighter elements. The isotopes initially formed are nearly all unstable and decay towards stable isotopes by the emission of beta-particles, accompanied in many cases by gamma-rays. A large range of radioactive half-lives is present in the mixture, from times less than a second to greater than 10 million years. The total amount of radioactivity in the mixture (disintegrations per second) decreases with time; that is, the rate falls by a factor of 10 for each seven-fold increase in the time since they were formed. This rate of decrease continues for about the first six months, after which the falling-off is more rapid.

To consider the possible effects of this residual radiation it is necessary to distinguish two cases: an airburst, by which is meant an explosion high enough above the surface so that the fireball does not touch the ground; and a surface, or near-surface, burst. In the case of an airburst, the debris rises rapidly from the point of the explosion into the upper atmosphere. The larger the explosion the higher it rises. This is accompanied by no condensible material except the relatively small mass of the weapon itself. As a result, when the normally solid materials cool and condense, the particles formed are very small, and these fall back towards the earth slowly, reaching the surface only after many days or weeks. In such a case the ultimate distribution of the debris on the surface—where it can interact with man—may be reasonably uniform over the hemisphere (northern or southern) in which the explosion occurred. A small part, however—possibly a quarter of the total—may even appear in the other hemisphere, starting about six months later.

An exception to this is possible in the case of a small explosion from which the debris passed under a layer of rain-forming clouds. Rain falling through the debris could wash out most of the radioactive particles and bring them to the surface in a small area and concentrated form. In general, however, from a series of airbursts the residual radioactivity would in effect be added to that in the natural background. The seriousness of the consequences would depend on the total fission yield contributing to the fallout.

In a surface burst a very large amount of material is entrained with the rising cloud, so that on condensing, many larger particles are formed. These will descend much more rapidly, and about half the radioactivity will fall back to the surface in the first 24 hours or so following the explosion. Assuming a steady wind at the altitude to which the debris rises, the area affected would be oval-shaped with its long axis in the downwind direction. The greater the wind speed, the greater the distance reached in a given time, but the smaller the concentration of radioactivity on the surface. The concentration will be highest near the scene of the explosion. It will decrease as one moves along the axis of the pattern and as one moves away from the axis. With a 15-knot wind, for example, at a point 60 miles from the burst, there would be no radioactivity until about four hours after the event. The deposit would then begin and build up over the next hour or so until the cloud passed by. With a variable wind the pattern would no longer be a smooth oval but might be highly irregular and difficult to predict.

There is no point at which the accelerated fallout from a surface burst actually stops, but the major portion of the effect is realized in the first 24 hours or so, and what comes down in that period is usually referred to as "local" fallout. The debris not involved in this local fallout will mostly be distributed worldwide, much as in the case of airbursts. The seriousness of the consequences at any particular point will be proportional to the intensity of the radioactivity deposited there, and this, as well as depending on the distance and the wind, will be determined by the fission yield of the explosion (or explosions) contributing to the fallout.

With respect to worldwide fallout, the United Nations Scientific Committee on the Effects of Atomic Radiations has studied the increase in radiation exposure caused by atmospheric tests. These tests, mainly those conducted by the United States and the Soviet Union prior to 1963, had about 200 megatons of fission yield. In the *1978 Yearbook* of the Stockholm International Peace Research Institute (SIPRI) the combined delivery capability of the United States and the Soviet Union was estimated to be about 15,000 megatons, of which something like 8,000 would be produced by fission. If all these weapons were fired in an all-out war, the residual radioactivity in the Northern hemisphere would be about 40 times larger than that resulting from testing and observed in 1964 and 1965.

The increase in the exposure over the 30 to 40 years following these explosions would be about twice that produced by natural background. There are many millions of people living in mountainous areas of the world, where they receive between two and three times as much natural background radiation as those living at sea level. Since there do not appear to be any large consequences from this exposure, one may conclude that this war-generated increase, though having some effect, would not in itself be disastrous for the whole population.

Such additional radiation would be expected to cause an increase in the cancer death rate of about 4 percent over the present spontaneous cancer death rate. The correlation between radiation exposure and the incidence of genetic disease has not actually been observed in humans. On the basis of observations made on animals, however, one could predict that the increase in significant genetic disease from this higher radiation level would be comparable to—or perhaps a little smaller than—the increase in the cancer death rate. The truly horrible effects of an all-out nuclear war would be those experienced in or near the target areas, rather than those in regions beyond the range of local fallout, apart, of course, from economic effects and the possible breakdown of existing social arrangements and standards of behavior.

Worldwide effects other than those from radiation could result from a large number of nuclear explosions, at least in principle. Examples include the propulsion of large quantities of dust into the atmosphere, and the injection of nitric oxide into the stratosphere, which would deplete the ozone layer. If pressed far enough, either of these processes could affect the heating of the upper atmosphere which, in turn, could lead to changes in the wind patterns at the surface and to climatological changes.

A large reduction in the ozone layer would lead to an increase in ultra-violet radiation reaching the Earth's surface, and hence to an increase in skin cancer in humans or, conceivably, to changes in the balance of natural biological systems. These processes are discussed extensively in the 1975 report of the National Academy of Sciences, *Long-term Worldwide Effects of Multiple Nuclear-*

Weapons Detonations. Many aspects of the mechanisms involved are adequately understood, but a number of the important interactions are not.

The conclusion of the Academy study was that some of the predicted effects that have been considered would, indeed, occur; that the changes induced would persist for only a rather few years; and that the magnitude of the changes would most probably be within the normal variability, in the case of climate, or—for the incidence of skin cancer—to an initial increase of no more than about 10 percent. These conclusions were supported to some extent by experience. The amount of dust put into the atmosphere was within the range estimated to have resulted from the eruption of Krakatoa in 1883, which led to a lowering of average temperature on the surface by a few tenths of a degree centigrade for a year or so. The nitric oxide effects seem likely to be limited, at least in part, by the fact that greater dependence on multiple reentry vehicles leads to smaller rather than larger yields in an increasing portion of the major stockpiles; only for yields of more than a megaton does the debris rise high enough to interact effectively with the ozone layer. These effects, then, are most probably not of great significance; but there were, at the time of the report, enough uncertainties in the estimates of some of the important processes so that it could not be absolutely established that the effects were in the moderate range.

The major effect still to be considered is that of "local" fallout. The natives on Rongelap Atoll—105 miles from the surface explosion of 15 megatons on Bikini Atoll on March 1, 1954—received about 175 rem before they were evacuated 48 hours following the explosion. None of these people was incapacitated as a result of the exposure. But all had anomalous blood counts throughout the next 15 years, and beginning about 10 years after the event there was a marked increase among the exposed group in the number of thyroid lesions requiring treatment. At exposure levels of about 450 rem, it is expected that essentially all persons would require hospital treatment, and that about 50 percent would die. At about 600 rem or above, most persons would die within a few days or weeks. The National Council on Radiation Protection has concluded that about 200 rem received in a short time marks the dividing line between exposures that will and will not cause sickness requiring medical care.

The local fallout from a surface burst of a one-megaton weapon would result in a patch of about 200 square miles (perhaps six miles wide and up to 45 miles long) on the edge of which the radiation exposure during the 24 hours following the explosion would be lethal to most persons remaining there without protection. Exposures would be higher at points within the contour. On a larger contour (say, 10 miles by 60, and enclosing about 450 square miles) the exposure in the first day would cause illness in most unprotected people.

To avoid death or injury the people in such a region would have to reach adequate shelter or evacuate the area within an hour or so after the fallout first arrived. A well-built house provides a reduction factor of only two or three, so it would not be enough merely to stay indoors, although a basement without windows or exposed walls would be expected to have a reduction factor about 10 times larger, which might be adequate in most cases.

Sheep or cattle left out on pasture would receive lethal exposures within a somewhat more extended contour, partly as a result of gamma-radiation from the fission fragments on the surface, but also as a result of ingesting contaminated grass. Some growing crops could be damaged and provide a reduced yield; and for certain mature crops, such as grain, there could be some hazard involved in harvesting them, in addition to the fact that the end product might be too contaminated to be acceptable in the food supply.

Considering the expected fairly rapid decrease in radioactivity, it would be possible to re-enter parts of such an area after a few days for some important purpose and for a limited period; but it would not be safe to resume normal activities for many months. Evidently, the effects of local fallout from a surface burst, though not so unavoidably lethal as the direct effects, are serious over much larger distances and times. The problems would be more complicated and severe if several surface bursts were to occur close enough together so that their fallout patterns overlapped.

An extreme example of this latter type would be provided by a full pre-emptive attack on the U.S. Minuteman force. With the great advances in guidance accuracy it is possible to consider targeting a single missile in its launching site. Since the missile silos are highly blast-resistant, large explosions would be needed; and to ensure a satisfactorily high probability of destroying the silo might require aiming two weapons at each. Also, a surface, or near-surface, burst would be indicated to provide the very high pressures (hundreds of pounds per square inch) needed to destroy the installation. The Minutemen are located in six "fields" in the mid-continent, each field being from 60 to 100 miles on a side and housing 150 or 200 Minutemen, spaced far enough apart so that they could only be attacked separately. To neutralize the force would require an attack designed to destroy all the 1,000 missiles in the force on a near-simultaneous basis.

In such an event rather few civilians would be within range of the direct effects of the attacking explosions. However, the debris clouds from the 300 or 400 explosions at each "field" would be close enough together to amalgamate and produce a 300- or 400-megaton fallout pattern. The area of serious hazard (exposure in the first 24 hours resulting in illness or death to unprotected persons) from one of these patterns would be about 70,000 square miles – perhaps about 600 miles long and maximum width of about 150 miles. There would be six such areas, but some of them would be close enough together for their patterns to merge.

From most locations within a few hundred miles downwind from the source, evacuation soon enough and reliably enough to avoid dangerous exposure would scarcely be possible, because of distances involved and uncertainty as to where safe areas might be found. With the field at Whiteman Air Force Base only about 160 miles to the west, the 2.5 million metropolitan-area residents of St. Louis would be within range of an exposure of about 3,000 rem in the first 24 hours. To avoid injury it would be necessary to reach a well-shielded location within an hour or less and remain there for a week or more. Depending on wind patterns, various other large cities could be in the path of fallout at dangerous levels, and some protective action might be required at points as far east as the Atlantic coast, though probably only in a rather narrow band. The farther from the source the more feasible evacuation would become, because of longer warning times, lower intensities and narrower fallout patterns.

Estimates of the number of fatalities that might result from such an event cover a wide range—from about 3 million to over 20 million. The only clear fact is that the number would be large.

In addition to the fatalities and injuries and the huge number of displaced persons, there would be a very serious disruption of manufacturing and distribution systems—whether for food, fuel, medical supplies, or any other essentials. There would also be the loss of livestock and the current year's crops from a very important fraction of the country's most productive agricultural land. The further question would be how much of this land could be put back into use for the production of foodstuffs in the year or so following, depending on the level of radioactive contaminants in the soil.

The events pictured may seem implausible. But they are implicit in the government-sponsored scenario in which the Minuteman force is judged to be so vulnerable that it is essential to deploy the MX in order to correct the situation.

What has been described is often referred to as "limited" nuclear war, which would not necessarily involve any direct attack on the civilian population—but that, of course, could follow.

9 Mechanics of fallout

BERNARD T. FELD

The process of fission in uranium or plutonium results in two effects. One is the release of a large quantity of energy—millions of times more, per fissioning atom, than that which is released in the chemical reactions involved in "ordinary" explosions; the second effect is the highly radioactive residue of fission products—two radioactive nuclei for each nucleus undergoing fission. Both these characteristics of the fission process are immutable: determined by the nature of the uranium and plutonium nuclei and by the forces which bind their constituents together. There is no known or foreseen way of altering these characteristics.

The fissioning of one kilogram (2.2 pounds) of uranium or plutonium produces:

- an energy release equivalent to the explosion of 20,000 tons of TNT;
- one kilogram of radioactive fission products, emitting mainly beta and gamma radiation (in the range of 0.1 to 1 million electron volts, or MeV) with the initial intensity of 67 thousand billion (6.7×10^{13}) curies.[1]

The fission products from the explosion decay with a great variety of lifetimes, each one characteristic of the radioactive fisson product nucleus or its successor in the radioactive decay chain. One hour after the explosion of 1 kilogram of uranium or plutonium, the radioactivity of the products is reduced to 3.6 billion curies. From one hour to about six months after the explosion, the radioactivity of the fission products may be approximated by the following simple (power law) expression

$$\frac{3.6 \times 10^9 \text{ curies.}}{t(\text{hrs})^{1.2}}$$

The total (integrated) radiation emission from 1 kilogram of fission products is around 100 billion (10^{11}) curie hours, of which one-half is emitted within the first minute and around four-fifths within the first hour.

As noted, there is a great variety of fission products. The danger to human health depends on the particular product under consideration, its propensity for incorporation into the biosphere, and the nature of its radioactivity. Table 1 lists the properties of three of the products that are among the most important from this point of view.

Although most fission products decay relatively rapidly, some

Table 1

Radioactive properties of three fission products of a nuclear explosion
There is a great variety of fission products and residue from a nuclear explosion. These three products and their properties are among the most important to consider when discussing the consequences of nuclear fallout.

Fission product	*Half-life (years)*	*Fission yield (%)*	*Decay products (MeV)*
$_{38}Sr^{90}$	28	5	0.54[a]
$_{53}I^{129}$	17×10^6	1	0.15[a] 0.04[b]
$_{55}Cs^{137}$	30	7	1.2[a] 0.66[b]

[a] Beta-ray energy
[b] Gamma-ray energy

(for example, iodine-129) have extremely long half-lives. However, in addition to fission products the debris from a nuclear detonation contains heavy nuclei (actinides)—some of the original plutonium which has not been entirely consumed in an explosion of less than perfect efficiency, and some products of neutron capture by the original plutonium or uranium. The actinides are predominantly alpha-ray emitters.[2] Because of their low-penetrating power, they are only harmful when ingested. However, when actinides are deposited in the lungs or the bone marrow, the potential risk for the eventual induction of cancer is increased. The main actinide products of a nuclear explosion and their properties are listed in Table 2.

Finally, the products of neutron-induced fission include additional neutrons—somewhat more than two, on the average, per fission. It is these excess neutrons that permit the continuation of the fission chain and make possible the explosion. However, the number of neutrons emitted is well in excess of the number required to perpetuate the chain reaction, and many of these excess

Table 2

Radioactive properties of actinide products of a nuclear explosion

In addition to fission products, the debris from a nuclear explosion contains many nuclei (or actinides), which are harmful when ingested.

Actinide	*Half-life (years)*	*Alpha-ray energy (MeV)*	*Gamma-ray energy (and yield) (MeV)*
$_{94}Pu^{239}$	24,000	5.15	0.05 (30%)
$_{94}Pu^{240}$	6,580	5.16	0.05 (25%)
$_{95}Am^{241}$	460	5.48	0.04 (15%)

neutrons escape into the biosphere. These neutrons represent a very important source of damage to individuals.[3] This is because of the effectiveness of these neutrons in producing damage to human tissue and organs and because of the lingering radioactivity resulting from their absorption by common materials in the surroundings. Some of the more important neutron absorption products and their radioactive properties are listed in Table 3.

The radioactive products and their distribution depend critically on the nature of the explosion (for example, whether it is in the atmosphere or on or near the ground) and on the meteorological conditions that prevail at the time (direction and intensity of wind, rainfall pattern, etc.). There is some experience with both kinds of explosions as a result of the extensive testing that took place above ground prior to the Partial Test-Ban Treaty of 1963

Table 3

Radioactive properties of neutrons absorbed by common materials after a nuclear explosion

In addition to fission products and actinides, the products of a nuclear explosion (or neutron-induced fission) also include additional neutrons. These neutrons permit continuation of fission chain and make possible the explosion. However, the excess neutrons which escape into the biosphere are an important source of damage to individuals.

Radioactive nucleus	*Half-life*	*Beta-ray energy (MeV)*	*Gamma-ray energy (MeV)*
$_1H^3$ (tritium)	12.26[a]	0.018	–
$_6C^{14}$	5,770[a]	0.156	–
$_{11}Na^{24}$	15[b]	1.39	2.75 1.37
$_{17}Cl^{36}$	3.1×10^5[a]	0.71	–

[a]years [b]hours

(see Table 4). The great bulk of the tests since 1963 have been underground, although there have been some French and Chinese nuclear test explosions above ground. Thus, aside from occasional, relatively minor "venting," the radioactivity from these underground tests has been confined.

For near-ground bursts, on the other hand, the main effects are directly downwind, arising from the radioactivity of the particles of dirt, etc., which are first carried aloft by the fireball and then slowly deposited from the drifting cloud of debris. As a rough rule of thumb, a one-megaton fission ground burst can spread lethal radiation (that is, 500 to 1,000 rem integrated dose) over an area of up to 2,000 square kilometers (somewhat less than 1,000 square miles).

Table 4

Average dose of radioactivity received from various sources by all individuals in the northern hemisphere[a] as a result of the approximately 1,000 megatons of fission explosion that took place in the atmosphere prior to 1963 partial test-ban treaty.

Source	*Approximate average dose (rem)*
${}_{6}C^{14}$	0.1[b]
${}_{55}Cs^{137}$	0.1
${}_{38}Sr^{90}$	0.5
total (whole body)	0.1-1
average from natural sources per year:	0.13

[a] Fallout in the southern hemisphere was measured to be about one-third of that in the northern hemisphere.

[b] The rem, or radiation equivalent man (a measure of biological "effectiveness"), is approximately equal to the rad for beta and gamma radiation, and up to 10 times greater for alpha rays or neutrons.

A nuclear war between the United States and the Soviet Union is likely to involve much more than one megaton of fission. (Since most strategic weapons in the superpower stockpiles are of the old-fashioned, dirty, fission-fusion-fission variety, and around half the energy yield is from fission, the distinction is not made here between explosive energy and fission energy.) If, for example, a so-called counterforce exchange occurred *today*—one or the other side, or both, attempted to eliminate the land-based strategic missile force of the other in a pre-emptive first strike—it would involve at least a few thousand megatons in a near-ground burst mode. Thus, a very large fraction of the area of the attacked nation would be covered by lethal fallout. (The area of the United States is around 5 million square miles and that of the Soviet Union about twice as large.

It is expected that, by the end of this decade, the available megatonage on both sides will have grown to some 10,000 to 20,000 megatons each. It is correspondingly not unlikely that the offensive deployments (MX, SS-18, 19) on both sides will have grown to accommodate this increase. A counterforce exchange of this magnitude would effectively eliminate the two superpowers as viable states, wiping out a major fraction of their populations as well as their accumulated wealth and industrial base, rendering their farmlands unusable, their waters undrinkable, etc.

Furthermore, such an exchange would guarantee a radioactive burden of some 10 to 40 rem to all the remaining inhabitants of the Earth (including those of the aggressor nation). The short-and long-term effects of this kind of universal exposure are not predictable. (The best current estimate predicts approximately one delayed casualty per 10,000 man-rems, but this presumably does not include longer-term genetic damage.) The human race might survive a nuclear exchange, but the genetic burden on future generations would be enormous.

Clearly, we are not so far from (and rapidly approaching) the situation depicted some 25 years ago by Nevil Shute. In his novel

On the Beach, Shute described the final hours of a small band of Australians on a planet deprived of the possibility of mammalian survival as a result of a nuclear war in the northern hemisphere. Today, it is possible to calculate the lethal quantity which would guarantee our permanent extinction:

1 beach = 1 million megatons of fission.

This is the size of a nuclear war that would inevitably bequeath the Earth to the cockroaches. It is sobering to note that current world stockpiles of enriched uranium and plutonium are already within a factor of less than 100 of this horrendous limit. An optimist, however, would rephrase this to read: "Current world stockpiles are not yet within a factor of 100 of this limit."

1. The curie is a unit of radioactivity. One curie is the radioactive decay rate of one gram of radium, or 37 billion (3.7×10^{10}) disintegrations per second.

The intensity of radiation depends on the geometrical distribution of the source and on the distance from the source to the irradiated object. (The intensity from a point source falls off as the square of the distance from the source.) The intensity is measured in radiation units or rad; 1 rad represents a flux of one billion (10^9) particles over an area of 1 square centimeter. 1 rad of gamma radiation of 1 MeV energy, on traversing normal tissue, results in the deposit of 100 ergs per gram of tissue. A source of intensity 1 curie at a distance of 1 meter yields 1 rad per hour. Recapitulating:

1 curie = 3.7×10^{10} disintegrations per second;
1 rad = 10^9 particles per square centimeter;
$\rightarrow$ 100 ergs per gram of tissue;
1 curie hour at 1 meter $\rightrightarrows$ rad.

2. The alpha particle is the nucleus of the helium atom $_2He^4$.

3. The neutrons that escape immediately into the biosphere are known as "prompt" neutrons. There are also some "delayed" neutrons. But while these play a crucial role in making possible a controlled chain reaction, they can be neglected for the purposes of this discussion.

10 Consequences of radioactive fallout

PATRICIA J. LINDOP and J. ROTBLAT

Radioactive fallout is a unique property of nuclear weapons, and there is no real experience of how to handle it. The bombs used on Hiroshima and Nagasaki in 1945 were detonated at such heights that no early fallout occurred, except for a "rainout" in a few localities. The atmospheric tests of nuclear weapons were generally carried out in uninhabited areas, so that the main effect on populations was that from global fallout. There are persistent rumors of military personnel and civilians having been exposed in the vicinity of tests, but no quantitative data have been released. The only documented event of an exposure of people to local radioactive fallout is the Bravo test of March 1, 1954, on the Bikini Atoll in the Pacific Ocean. Although the number of persons exposed was small, this incident provided valuable material for estimating the extent and, particularly, the duration of the effects of fallout. However, it hardly gives an idea of the magnitude of the problems which would face the medical profession following the detonation of a large number of nuclear weapons in a densely populated country.

Radiation doses from fallout

The radioactivity in the fallout can expose populations in several ways, and in different time sequences:

- external irradiation by the radioactive cloud as it passes overhead;
- internal irradiation through the inhalation of radioactive particles in the air;
- external irradiation, mainly by the gamma-rays from the

radioactive substances deposited on the ground;

• internal irradiation through eating meat or drinking milk from animals which had ingested radioactive substances, or by eating food from plants which had incorporated such substances, or by drinking contaminated water.

In the case of local fallout the external irradiation by gamma-ray exposure from matter deposited on the ground represents the most important hazard. It gives rise to total-body exposure, and the dose rate is proportional to the deposited activity. If all the gamma-ray radioactivity resulting from the detonation of a 1-kiloton fission bomb were deposited uniformly over an area of 1 square kilometer, then the dose rate—at a height of 3 feet above ground—would be 7,500 rads per hour, one hour after the explosion. Allowing for unevenness of the terrain, which may cause some of the gamma-rays to be absorbed in the ground, and for the fact that only about 60 percent of the radioactive content of the fallout is deposited as local fallout, the gamma-ray radioactivity from the fallout of a 1-kiloton fission bomb would give rise to a dose rate of about 3,000 rads per hour at one hour after the explosion. For weapons of other yields, the dose rate increases in proportion to the fission content of the bomb.

The actual dose rate is modified by two factors: the decay of the radioactivity with time, and the spreading of the fallout with distance.

Time variation of dose rate and accumulated dose. The fission products undergo radioactive decay, and therefore the dose rate decreases rapidly with time (see Chapter 9). Table 1 gives the average dose rates due to the gamma-rays from early or local fallout, at different times after an explosion. The dose rates in the table are with reference to the dose rate at one hour after the explosion, at which time it is assumed to be 100 rads per hour. If in any actual situation the dose rate at one hour, at any other time, is known, then the dose rates at other times can be obtained simply by proportion.

It should be noted that the variation of dose rate with time, as given in Table 1, is valid only after the fallout is complete at a given place, and if no additional debris or material is brought into the location by another explosion and if no material is dissipated because of weather.

The total dose received by a person is obtained by multiplying the dose rate by time of exposure. Since, however, the dose rate rapidly decreases with time, the calculation of the total dose involves an integration over the relevant values of dose rates. The result of such integration is presented in Figure 1 which gives the accumulated dose (in rads) starting from one minute after the explosion up to the given time, assuming that the dose rate at one hour is 100 rads per hour. For other values of the dose rate, the values in Figure 1 have to be multiplied by the actual dose rate at one hour.

Figure 1 also allows the calculation of the total dose received by

Table 1

Dose rates due to gamma-rays from fallout at various times after a nuclear explosion

Time (hours)	*Relative dose rate (rads per hour)*
1	100
2	40
4	15
6	10
12	5.0
24	2.4
36	1.6
48	1.1
72	0.62
100	0.36
200	0.17
500	0.050
1,000	0.023

The radioactivity of the fission products decreases with time, resulting in a corresponding decrease of the dose rate. The table shows the dose rates at various times after the explosion, if at one hour the dose rate was 100 rads per hour.

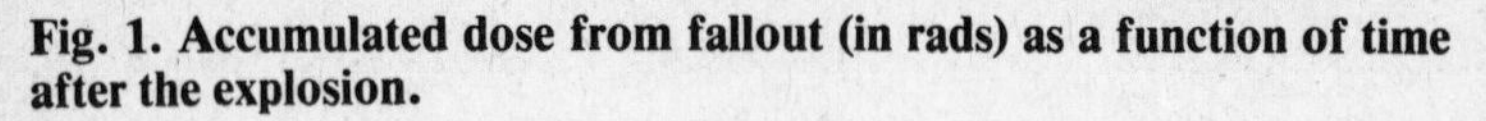

Fig. 1. Accumulated dose from fallout (in rads) as a function of time after the explosion.

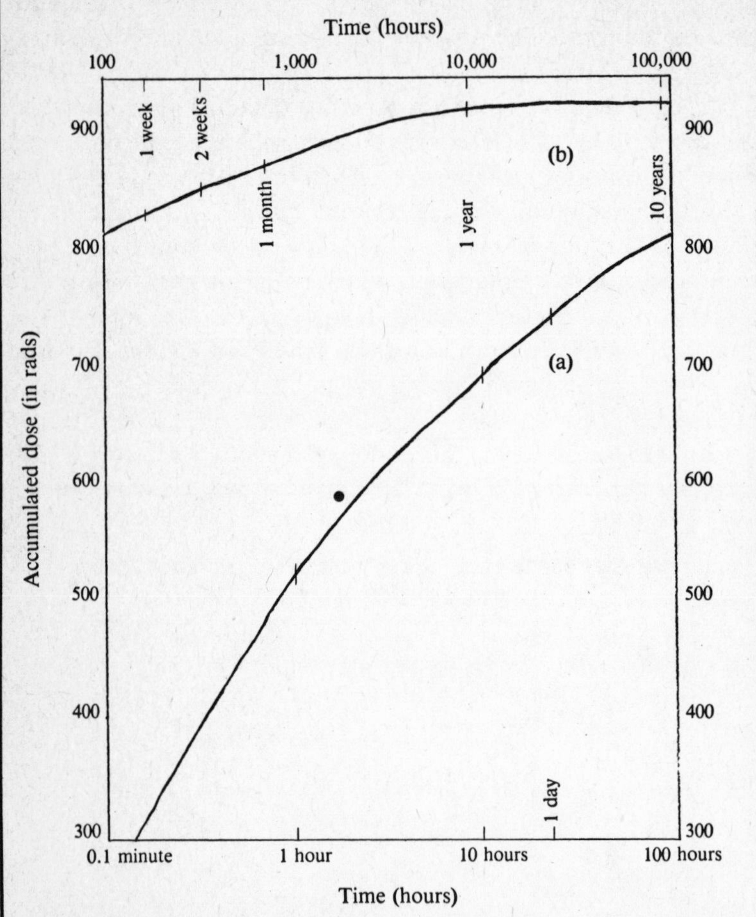

The total dose received by a person is obtained by multiplying the dose rate by the time of exposure. As the dose rate decreases with time, the calculation of the total dose involves an integration over the relevant values of dose rates. The result of such integration is presented in Figure 1.

The figure assumes that the dose rate is 100 rads per hour at one hour after the explosion. For times up to 100 hours, use curve (a); for times longer than 100 hours use curve (b).

a person who enters a given fallout locality at a certain time after the explosion and remains in it for a certain period. This dose is given by the difference in the ordinates corresponding to the times of entry and exit. As is seen from Figure 1, the total accumulated dose tends toward a finite limit, namely 930 rads if the dose rate at one hour is 100 rads per hour. This value, 930 rads, gives the infinity dose, that is, the total dose accumulated starting from one minute after the explosion until an infinite time.

The method of dose calculation outlined above applies only to a single detonation. If the fallout in the given locality is caused by several bombs, exploded at different times, the variation of dose rate with time will be quite different from that given in Table 1, and a single measurement of the dose rate in that locality would not be sufficient to calculate the accumulated dose.

The rapid rate of decay of the early fallout is often used to reassure the population that the danger from fallout radioactivity is over after about two weeks. This is misleading. What matters is not the dose rate but the accumulated dose, and the latter decreases with time less rapidly than the dose rate. In any case, if the initial fallout in the given area was of high intensity, then the exposure level could be dangerously high for a long time and the areas may remain uninhabitable for many years. For example, if the dose rate at one hour was 10,000 rads per hour, then a person entering the fallout area after one month and then remaining there could accumulate a dose of about 5,000 rads in one year. Even if he entered after one year, the dose accumulated during the next year could be about 300 rads.

Distribution of "early" fallout. For a given bomb the distance traveled by the fallout particles, and the time and location of their deposition, are primarily dependent on the speed and direction of the wind. As the particles are carried further away by the wind they spread over larger areas so that the rate at which the radiation dose may be delivered is rapidly reduced with the distance. This decrease is in addition to the radioactive decay which occurs during the time before the fallout particles reach the ground.

Fig. 2. Contours for fallout from a 2-megaton bomb at 18 hours after the explosion: (a) dose rates and (b) total dose

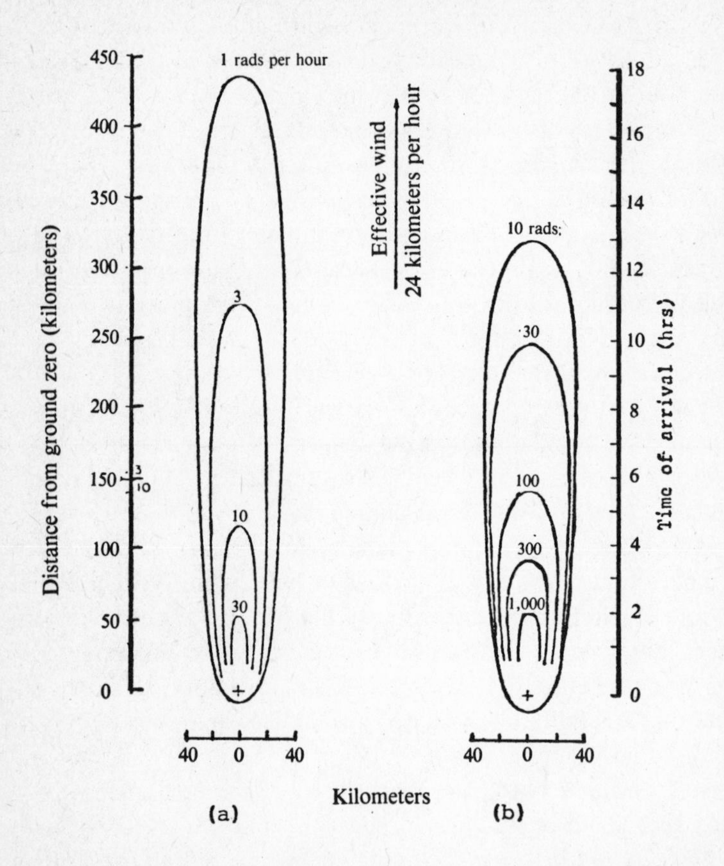

Under ideal conditions the lines joining all points with the same dose rate, or the same dose, at a given time, are cigar-shaped. The figure shows several such contours for the fallout for a 2-megaton bomb with a 50 percent fission content at 18 hours after the explosion with a wind velocity of 24 kilometers per hour: (a) shows dose rate contours; (b) shows accumulated dose contours.

Under steady wind conditions, with no shear, at any given time after the explosion, the radioactivity will have spread itself in such a way that the lines joining all points with the same dose rate will be cigar-shaped. Figure 2 shows (a) several dose-rate contours for the fallout from a 2-megaton bomb with a 50 percent fission content at 18 hours after the explosion, when the wind velocity was 24 kilometers per hour; and (b) the contours of the total doses which would result from exposure at these dose rates.

It should be noted that these contours represent a transient situation in space and time. At any given place the activity first increases with time, as more fallout reaches a locality and is deposited, but subsequently it decreases, due to natural decay.

Figure 2 is intended to illustrate the situation in a grossly simplified manner only. In reality the picture is much more complex because many factors may disturb the pattern of the fallout. First, the wind may change in speed or in direction, and this would distort the shape of the contours. Then, there may be scavenging of the fallout by rain or snow, if the air-borne particles encounter a region where precipitation is occurring. Such events may give rise to "hot-spots," small areas where a very large amount of radioactivity is deposited. Rain coming down after the fallout may wash away some of the activity. All these phenomena have actually been observed in test explosions.

It is clear that the prediction of the hazard due to early fallout is highly unreliable. For example, Figure 3 shows a comparison of an idealized fallout pattern with a pattern that might actually result from variations in local meteorological and surface conditions. Nevertheless, an idealized picture is still useful in providing an approximate estimate of the possible radiation hazard from early fallout. Assuming a wind of constant velocity with little directional shear, and no precipitation, the parameters of the fallout pattern can be computed for bombs of different yields exploded near the surface.

Table 2 contains results of such calculations for 1-megaton and

Fig. 3. Idealized (a) and possible (b) dose-rates for a 10-megaton surface burst and a wind velocity of 50 kilometers per hour

(a) Idealized dose-rate contours (rads per hour)

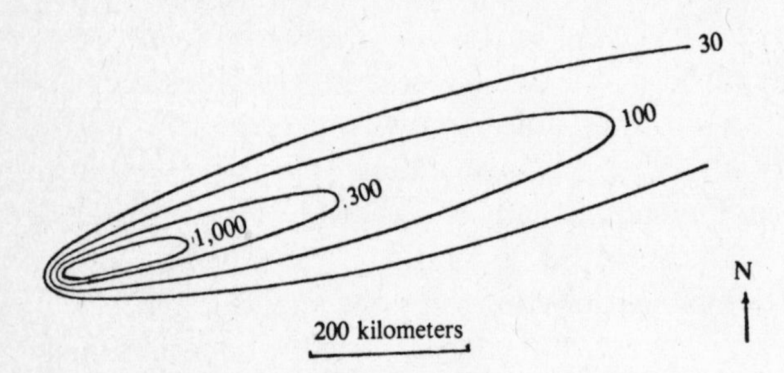

(b) Possible dose-rate contours (rads per hour)

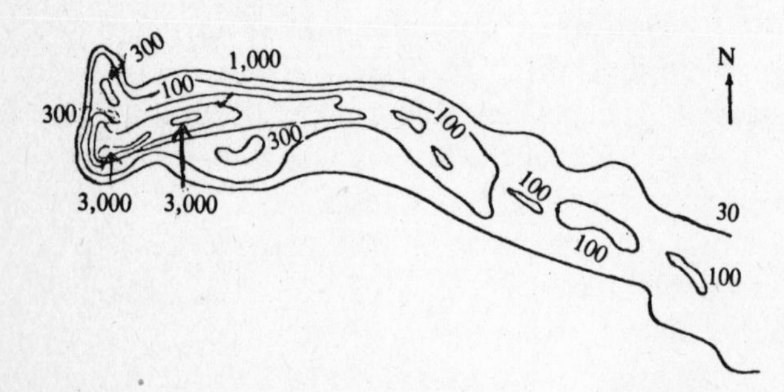

The actual distribution of fallout depends on many, mostly unpredictable factors. The figure shows dose rate contours for the fallout from a 10-megaton bomb with a presumed wind velocity of 50 kilometers per hour: (a) shows an idealized pattern; (b) shows the actual contours, distorted due to variations in local meteorological and surface conditions.

Table 2

Dose rates and accumulated doses in an idealized pattern of fallout

Downwind distance (km)	*Time of arrival (hrs)*	*Reference dose rate (rads/hr)*	*Accumulated dose (rads)*
	1-megaton bomb		
100	3.3	270	850
150	5.4	160	440
200	7.5	110	280
250	9.6	76	180
300	11.7	54	120
350	13.7	42	92
400	15.8	32	67
450	17.9	25	51
500	20.0	20	39
550	22.1	16	30
600	24.2	13	24
650	26.2	12	21
700	28.3	9.0	16
750	30.4	7.4	13
800	32.5	6.3	10
	10-megaton bomb		
100	2.8	1410	4570
150	4.9	670	1870
200	6.9	450	1160
250	9.0	330	800
300	11.1	260	600
350	13.2	220	480
400	15.3	180	380
450	17.3	160	320
500	19.4	130	260
550	21.5	110	220
600	23.6	98	180
650	25.7	87	160
700	27.8	74	130
750	29.8	66	110
800	31.9	61	100

Under idealized conditions, one can calculate the various parameters of the fallout from a bomb of given explosive yield. The table gives several of these parameters as a function of the distance from the explosion.

10-megaton thermonuclear bombs with half of the yield due to fission. The time of arrival of the fallout, the reference dose-rates (at one hour after the explosion), and the accumulated infinity doses for persons in the open are presented as a function of downwind distance. (The wind velocity was assumed to be 24 kilometers per hour.) If the idealized fallout contours are considered as ellipses, then areas can be calculated within which given total doses could be accumulated. The area seems to be a more suitable parameter than distance, since even if meteorological conditions distort the contours, the area covered by the fallout is likely to be less affected.

Table 3 shows the result of such area calculations for the 1-megaton and 10-megaton explosions considered in Table 2. It gives the areas within which the total accumulated dose could reach the value given in the first column. The areas were calcu-

Table 3

Areas covered by given accumulated doses from fallout

Upper limit of accumulated dose (rads)	*Area (square kilometers)*	
	1-Mt bomb	*10-Mt bomb*
1,000	1,000	12,000
800	1,300	16,000
600	1,700	21,000
400	2,600	29,000
200	5,500	52,000
100	10,500	89,000
50	18,600	148,000
25	32,700	234,000
10	56,000	414,000

For the same conditions as in Table 2, Table 3 gives the areas within which the maximum accumulated dose could reach given values. Thus, for a 10-megaton bomb persons in an area of 21,000 square kilometers might receive a total dose of 600 rads. Higher doses might be received in smaller sections of that area.

lated down to a value of 10 rads, which is approximately the limit of the dose which populations may accumulate in a lifetime from peace-time activities involving exposure to radiation (apart from the natural background and medical procedures). It is seen that one 10-megaton bomb could produce a dose near this limit over an area which exceeds the land area of almost every European country.

Factors affecting dose. The estimates made in the previous section are theoretical values, and represent the maximum gamma-ray doses that a person might receive from local fallout. In practice, the gamma-ray doses are likely to be smaller. Because of the long delay in arrival of the fallout in localities remote from the explosion, people would have time to leave the area threatened with fallout, or take shelter. The former is based on the assumption that uncontaminated areas would be left and that they could be identified. This may be feasible in the case of one or a few bombs, but is unlikely in the aftermath of massive bombing.

Even though properly designed shelters are unlikely to be available for the majority of the population, people may be expected to stay indoors, at least during the early period after the bombing. This again assumes that houses fit for people to live in would be left. Staying indoors would reduce considerably the gamma-ray dose. The dose reduction factor depends among other things on the type of the building, the floor level in a multi-story building, and its proximity to other buildings. On the average, a reduction by a factor of 5 can be assumed. This moderation of the hazard would of course not apply to domestic animals left in the open, or to crops.

On the other hand, there are factors which may bring about an increase in the dose and in its biological effect. The dose estimates made previously did not take into account the effects of the beta-rays, which are emitted by nearly all fission products. Beta-rays can contribute to the *external* hazard, if the radioactive materials come into direct contact with the skin, or mucosa of the mouth and nose, or the eyes. Owing to the short range of beta-rays (a few millimeters in tissue), their action is confined to the superficial layers of the skin, but they may cause beta-burns. These start with itching and a burning sensation, and then may develop into weeping, ulcerated lesions, causing much discomfort. These lesions could also temporarily incapacitate potential medical helpers.

The main effect of beta-rays is as an *internal* hazard, which ensues when a person inhales air containing radioactive particles, or ingests such particles with contaminated food or drink. The effects due to inhalation depend markedly on the size of the fallout particles. The nose filters out large particles, 5 micrometers or more in diameter. It is these particles that descend first and give rise to the highest external fallout doses. For this reason, inhalation contributes relatively little to the hazard. But the smaller particles which do reach the lung may settle down not only in that organ but also, depending on their chemical form, in other organs, in the bones, liver, thyroid gland and so on.

Of greater importance is the beta-ray dose delivered via the ingestion route. While the whole-body dose resulting from the internal deposition of radioactive nuclides is a small fraction of the external dose, the doses to individual organs may be as large or even larger than from the external gamma-rays. For example, in the case of adults in the Rongelap Atoll, the radioactive iodine taken in with the contaminated water gave rise to an internal dose to the thyroid gland nearly the same as the external gamma-ray dose; in children the dose to the thyroid was 3 to 8 times larger than the whole-body dose.

The main problem about water, particularly in towns, would be its availability. Supplies are likely to be interrupted by the bombing, through the destruction of equipment, pumping facilities and pipes. This means that even if reservoirs were not contaminated the water would either leak out or would not be available. Where water supplies are not interrupted, radioactive contamination is not likely to be a limiting factor. Lack of piped water, however, would compel people to seek other sources. Rainwater stored in open cisterns in fallout areas will be highly radioactive and unsuitable for drinking until the fallout particles had settled to the bottom. Rain falling on land undergoes a natural process of purification, so that ground water will be much less contaminated.

Fallout may contaminate food when radioactive rain or dust settles on vegetation. The contamination can be removed by washing the food with clean water, if it is available. However, most of the contamination will come through direct assimilation of particles deposited on leaves and shoots of plants and this cannot be easily removed, except by waiting for the short-lived radioactive products to decay. The same applies to milk and meat from animals which have grazed on contaminated grass.

An overall assessment of the effect of internal exposure indicates that for local fallout the hazard is small compared with that from external exposure to gamma-rays as far as acute effects are concerned. It should be noted, however, that evidence from animal experiments implies that the combination of internal and external exposures may act synergistically, and the lethal dose may be reduced much more than would be expected from the separate effects of the internal and external exposures.

Radiation effects

It is generally accepted that any exposure to ionizing radiation may produce harmful effects; but the type and the severity of the effect, and the time of its appearance, vary considerably. It depends primarily on the dose of the radiation. With high doses the symptoms of exposure, which may lead to death, are noticed shortly after the exposure. These are known as acute effects, and have to be distinguished from the long-term effects, which may follow exposures to medium-high and to low doses, and may take different forms, the most prominent being the induction of cancer.

The biological effects of radiation have been the subject of intense study for many years, and it has been claimed that more is known about radiation hazards than about all other environmental or occupational hazards of modern society. Yet, when it comes to the quantitative estimate of the harm to man from a given dose of radiation, there are very large uncertainties, sometimes amounting to an order of magnitude. This applies to both acute and long-term effects.

Doses to organs in the body. A person may be exposed either externally (when the source of radiation is outside the body) or internally (when he or she inhales or ingests a radioactive substance). In the case of external exposure, it is important to distinguish between the amount of radiation reaching the surface of the body and that reaching the various organs inside the body. These two quantities differ because of the attenuation of the intensity of the radiation in passing through the body. The degree of attenuation depends on the properties of the radiation as well as on the size of the body and the depth of the given organ.

When considering the injury caused by exposure to radiation it is the dose to the given organ which is the determining quantity. However, the radiation doses which are usually quoted in descriptions of the effects of nuclear explosions refer to the tissue dose measured at the surface of the body. This also applies to the previous section of this chapter.

The distinction between surface doses and organ (or midline tissue) doses is important when the population exposed consists of individuals of different sizes, including children. For example, for the same level of radiation, an infant will receive a larger dose to the bone marrow than an adult. When combined with the greater intrinsic sensitivity of children to radiation, this can literally mean the difference between life and death. A level of external radiation exposure which gives an adult a reasonable chance of survival is likely to kill a child. It is the babies and infants who will be the first to die after a nuclear attack.

Acute effects. The symptoms of acute effects appear soon after the exposure to radiation, within one hour or two (or even within minutes after very high doses); but death, if it is the result of the exposure, may not occur for some time. It is generally believed that death from acute effects comes within two months after exposure; but there is evidence from the bombing in Japan that it may come much later, up to several years. By that time deaths from long-term effects (particularly leukemia) begin to occur.

Acute effects may manifest themselves when the whole body, or a large part of it, is exposed in a short time to midline tissue doses from a few tens of rads upwards. The early symptoms, such as anorexia, nausea and headaches, are part of the so-called prodromal syndrome. With doses up to 100 rads these symptoms soon disappear and recovery is apparently complete. With increasing dose mortality increases, reaching 100 percent for a dose of about 500 rads to the marrow, although in healthy adults survival is possible even for larger doses if special treatment is provided. Death in the dose range of 100 to 500 rads is mainly due to damage to the blood-forming organs. Larger doses are invariably fatal, death being due to disturbances of the gastrointestinal system and homeostasis. At still higher doses the central nervous system fails.

Radiation sickness. Soon after exposure to radiation, a person may begin to show symptoms of acute gastrointestinal and neuromuscular effects. This prodromal syndrome is popularly known as radiation sickness. The gastrointestinal symptoms are anorexia, nausea, vomiting, diarrhea, intestinal cramps, salivation, dehydration and loss of weight. The neuromuscular symptoms include easy fatigability, apathy or listlessness, sweating, fever, headache and hypotension followed by hypotensive shock. All these symptoms occur only at high doses; with low doses, and during the first 48 hours after the exposure, only some of the symptoms may occur.

Many of the above symptoms may also be caused by factors other than exposure to radiation. The state of nervous tension when war breaks out, or crowding in shelters, may simulate radiation sickness. Witnessing genuine symptoms of radiation exposure in others may evoke similar symptoms, with epidemic results, particularly in young children. Without a physical measurement of the radiation dose received by a person it would be very difficult to distinguish between genuine and spurious reactions.

But even if the dose of radiation were known, it would be impossible in many cases to predict the outcome of the exposure in any individual. Very little is known about the mechanism which underlies the occurrence of the prodromal syndrome, but there is evidence of a large variation among people in their response to an exposure. This applies to all radiation effects. It has been suggested that in a given population there exist sub-groups with a higher than average sensitivity to radiation. The reason for the different sensitivities is unknown; it may include the genetic make-up and the general state of health. In any case, when many people are exposed to the same dose of radiation—in the lower range of the prodromal syndrome—some will show early symptoms of radiation sickness, but others will not. Therefore, in order to calculate the probability of a given symptom occurring, one has to take a statistical approach in terms of the percentage of an exposed population which will exhibit the symptom.

Table 4 gives the midline tissue doses for a 10, 50 and 90 percent probability of occurrence of various prodromal symptoms.[1] A dose of 50 rads is likely, for example, to induce nausea in 10 percent of persons exposed; a dose of 215 rads will cause vomiting in half of them; and a dose of 390 rads will cause diarrhea in 90 percent of those exposed as well as all the other symptoms.

The sequence of the symptoms, and the time of their onset and severity, may provide some guidance to the dose (if unknown) and to the prognosis. The higher the dose the earlier the onset and the more severe the symptoms. At moderate doses the symptoms subside after a few days, and the person exposed may then feel quite well for a week or so. But subsequently they reappear and together with other signs of malaise may lead to death, if the dose was high enough. Protracted vomiting during the first two days and, in particular, the occurrence of diarrhea indicate a bad prognosis. Persons with intractable nausea, vomiting and diarrhea will most likely die even if treated.

Median lethal dose (LD-50). In the dose range of 100 to 500 rads to the bone marrow the prodromal syndrome is followed by other clinical symptoms if too many bone marrow stem cells have been destroyed. These symptoms are hemorrhage under the skin, bleeding in the mouth and bleeding into internal organs. There is greater susceptibility to infection which causes a step-wise rise in temperature; severe emaciation and delirium lead to death, usually within six weeks. As in the case of the prodromal syndrome, the response to exposure to a given dose, within the range of 100 to 500 rads, differs from individual to individual. Some will die; but others will survive, and after some months the number of blood cells will be back to normal, and they will show an apparent recovery (although still at risk from long-term effects). In many cases recovery will take place after a long time and after possible

Table 4

Radiation doses to midline tissue (in rads) which produce radiation sickness symptoms

	Percentage of exposed population		
Symptom	*10%*	*50%*	*90%*
Anorexia	40 rads	100 rads	240 rads
Nausea	50	170	320
Vomiting	60	215	380
Diarrhea	90	240	390

A dose, for example, of 50 rads is apt to induce nausea in 10 percent of the individual exposed; a dose of 215 rads will cause vomiting in half of them; and a dose of 390 rads will cause diarrhea in 90 percent of those exposed.

occurrence of such diseases as pancreatitis and liver function changes.

To calculate the probability of death (or survival) the statistical approach is again necessary, but the amount of factual evidence available is very small. Figure 4 shows the percentage of persons, out of a large number exposed to a given dose of radiation, who are likely to die as a result of such an exposure.[2] The mid-point of the curve—the dose which gives a 50 percent probability of death

Fig. 4. Probability of death from acute effects

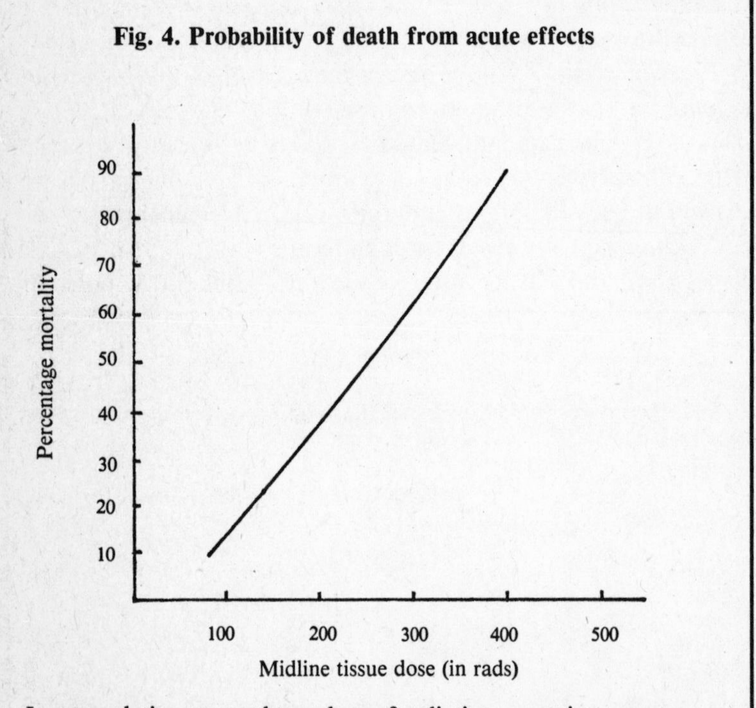

In a population exposed to a dose of radiation a certain percentage will die within a few months. The figure shows the percentage mortality as a function of the dose, to the bone marrow (taken as the dose to tissues at the midline of the body). The dose which results in 50 percent mortality is called the LD-50.

within a few weeks after exposure—is called the LD-50 dose (median lethal dose). It is an important parameter. LD-50 values have been measured carefully for many mammals and other living organisms, but for man it has been deduced from a small number of observations. The value in Figure 4, about 250 rads to the marrow, may therefore be subject to a large error, and there are suggestions that it is considerably higher.

A remarkable feature of the curve is its steepness. Within a very narrow range of doses, from 90 to 400 rads, the probability of death increases from 10 to 90 percent. The error in measuring the dose nearly covers this range, so that it would be difficult to say who received a lethal dose. Combined with the uncertainties in the LD-50 value and in the slope of the curve, this makes any estimate of the number of survivors after an exposure to a bone marrow dose in the range of 100 to 500 rads very dubious.

It should be noted that Figure 4 applies to adults who received whole-body exposure and who, though not receiving special treatment, would be under care, particularly to avoid infection after the exposure. In conditions of nuclear war this is unlikely to be the case, and the whole curve on Figure 4 might then shift to the left. The extent of the shift is impossible to predict, but any exposure above 100 rads to the marrow might result in death under these circumstances.

Factors affecting survival after acute exposure. Like the absolute value of the LD-50, the influence of the physical and biological factors on this value is not known sufficiently to be expressed quantitatively with any accuracy.

The dose rate affects markedly both the occurrence of the prodromal syndrome and the LD-50 value. From radiotherapy experience it is known that protraction of the exposure, either by giving it at a low rate or by dividing the dose into a number of fractions separated in time, requires an increase in the total dose in order to produce the same effect as a dose given in a short time. This indicates that the radiation damage is to some extent repaired in the time between the fractions, or during the protracted exposure. In the case of continuous exposure, observations indicate that if the total dose is given in eight days instead of in one day, then all the values in Table 4 for the prodromal symptoms would have to be doubled.

With regard to the LD-50 value, the curve on Figure 4 can be applied if the total dose is delivered within one day or less. If the same dose is spread over two weeks (the main period of exposure from local fallout), the LD-50 would be nearly doubled.

Of the biological factors, the most important is the proportion of the volume of the body exposed to the radiation. Parts of the body (for example, limbs) can be exposed to much higher doses before the acute symptoms occur. In total-body exposure an increase of the LD-50 value would also ensue if part of the body were shielded from the radiation or received a smaller dose. The greater the proportion of bone marrow screened from radiation the greater the chance of survival.

Age at exposure is another relevant factor. The sensitivity to radiation appears to be greater both in the very young and the old. For these population groups, the LD-50 value is likely to be smaller than for adults in the middle age group, but the amount of the decrease is unknown.

Acute lung effects. Death from the acute effects of radiation may also occur as a result of internal exposure, following the inhalation or ingestion of radioactive substances, for example, from fallout. Inhalation is of particular importance in its effects on the lungs.

If the dose to the lung tissue is high enough, the inhalation of radioactive particles may produce acute effects, even leading to death. This is quite apart from long-term effects such as fibrosis and cancer of the lung, which may result from much smaller doses. The acute effects ensue both from the direct action of the radiation on the lung walls and from the damaging action on the cells of the lung. The former affects the permeability of the membrane of the alveoli (air-sacs of the lungs), allowing fluids to escape into them. The symptoms of this are coughing, shortness of breath, and a feeling of "drowning" in lung fluids. The effects on the lung cells include swelling of the alveolar walls, resulting in reduced gas exchange with subsequent hypoxia (deficiency of oxygen); hemorrhage into the alveolar spaces, giving rise to blood-stained sputum; loss of surface-acting secretion, leading to collapse of the alveoli and consolidation of the lung; and loss of the immunological function of the lung, leading to infection and pneumonia.

The causes of death may be heart failure due to hypoxia, pneumonia, or generalized toxemia. The time of death may be some months after the inhalation; it depends on age, environmental conditions and availability of treatment. The lethal dose to the lungs is about 1,000 to 2,000 rads. A person inhaling a radioactive material is also likely, however, to have internal exposure to other organs as well as external exposure. The combined, possibly synergistic, action of the individual exposures makes the prognosis much worse. Under such circumstances smaller doses to the lung may prove lethal.

Estimates of casualties.

Quite large doses of radiation from fallout, in the range of acute lethal effects, could be accumulated at considerable distances from the detonation (see Table 2). Taking into account the variation of the LD-50 dose with the dose rate and the time distribution of the dose, and converting from bone marrow doses to surface tissue doses, one can calculate the distance at which a person in the open could accumulate an LD-50 dose. For the 1-megaton bomb this distance comes out to be about 125 kilometers; for the 10-megaton bomb it is nearly 300 kilometers.

From the data in Table 3 the lethal area for acute radiation effects (the area within which the number of survivors would equal the number of fatalities outside the area) can be calculated. For the 1-megaton bomb the lethal area is about 1,700 square kilometers; for the 10-megaton bomb it is about 20,000 square kilometers. The latter is nearly 400 times greater than the lethal area from the initial radiations from the bomb; it is also many times larger than the lethal areas for blast and heat effects.

If the population density in the fallout area is known, one can calculate the number of persons exposed to different doses, and the possible casualties. Assuming a population density of 100 persons per square kilometer (the average for Europe), then about 2 million persons might receive a dose from which they would die within a few weeks or months, if they were exposed to fallout in the open from a single 10-megaton bomb.

Finally, the effect of war conditions on the chances of recovery from an acute radiation injury must be taken into account. The acute mortality rates, calculated in the previous section, were based on LD-50 values applicable in normal conditions. With the lack of food, medical care and other social amenities, which is bound to occur in the wake of a nuclear war, many more people are likely to die following exposure to sub-lethal doses of radiation. It is difficult to envisage that people will stop eating food contaminated with radioactivity, which they cannot see or smell, when there is nothing else to eat.

Fallout from Bikini

An unintended exposure of a population to the radiation from local fallout occurred in the Marshall Islands, following the Bravo test of March 1, 1954. This was the first test of a large thermonuclear device, and the explosive yield was about 15 megatons. The detonation was near the ground, about 2 meters above a coral reef in the Bikini Atoll. Some of the radioactive cloud came down unexpectedly in a long plume in an easterly direction, covering the Marshall Islands, several of which were inhabited by natives and one by U.S. personnel. An area of about 20,000 square kilometers was contaminated to such an extent that lethal doses would have

Fig. 5. Dose contours after the Bravo test explosion

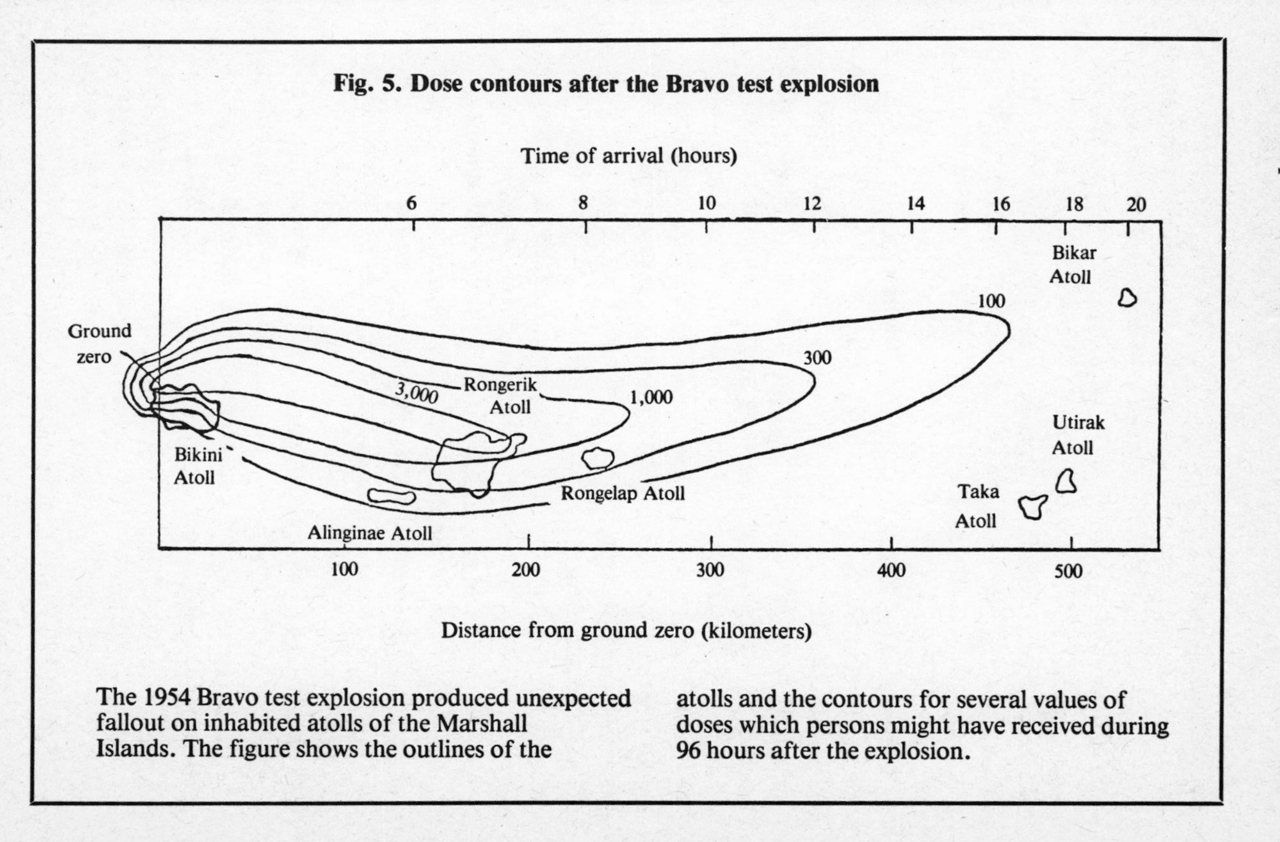

The 1954 Bravo test explosion produced unexpected fallout on inhabited atolls of the Marshall Islands. The figure shows the outlines of the atolls and the contours for several values of doses which persons might have received during 96 hours after the explosion.

been received by persons staying in the open. Figure 5 shows the outlines of the various atolls as well as contour lines for several values of total doses which persons might have received during the 96 hours after the explosion. (The accumulated doses to infinity would be nearly double those values.) The bottom and top scales give the distance from ground zero and the time of arrival of the fallout at the given location. Due to an insufficient number of monitoring instruments, the contours were drawn largely by guesswork.[3]

Two days after the test the inhabitants were evacuated. By that time some had received whole-body doses of up to 200 rads (to tissue at the surface), and the majority of the islanders on Rongelap Atoll had burns from beta radiation. There was also internal exposure due to inhalation and ingestion of radioactive materials (particularly radioactive iodine) with food and water. Some inhabitants exhibited symptoms of acute exposure (anorexia, nausea and vomiting). Long-term effects (predominantly thyroid disorders) appeared later. Almost all the children of the Rongelap Atoll had lesions of the thyroid and had to undergo surgery for the removal of thyroid nodules; later, a number of them showed symptoms of hypothyroidism. Several cases of cancer of the thyroid occurred among the female inhabitants.

The population of the Rongelap Atoll islands were allowed to return in 1957, 3 years after the test; but more than 20 years later, in 1979, the northern islands of the atoll were still declared to be too radioactive to visit.[4]

The islands of the Bikini Atoll, where testing continued until 1958, remained uninhabited for many years. Vigorous decontamination measures were taken, including the removal of 5 centimeters of the topsoil, before planting new trees. In 1967, the Bikini Atoll was declared habitable, and some islanders returned. However, further geological surveys showed that the activity in the soil was still too high for agriculture and the atoll was again evacuated. By the end of 1980 some islands of the atoll were declared safe for habitation, but only if 50 percent of the food for the inhabitants was imported.[5]

In addition to the inhabitants, a Japanese fishing boat was showered with fallout particles. The crew was exposed externally from the fallout deposited on walls and floors of the vessel and from the particles adhering to the surfaces of their bodies (see Figure 6), and internally from the incorporated radioactive materials.[6]

Fig. 6. Contamination and lesions of body surface

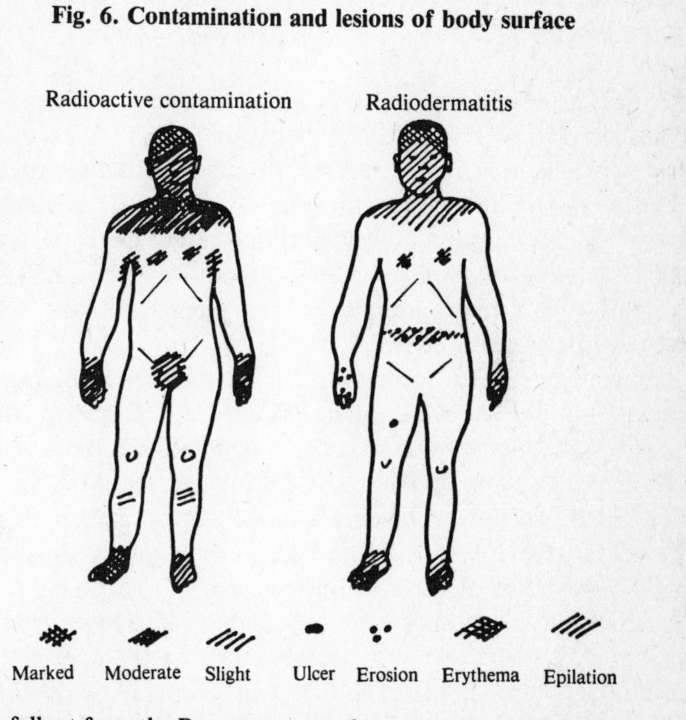

The fallout from the Bravo test came down on a Japanese fishing boat contaminating the crew with radioactivity. The figure shows the areas of contamination and the skin lesions resulting from the exposure to beta rays.

On the first evening their eyes were affected; they suffered from lacrimation and pain in the eyeballs. After two weeks, they developed photophobia, edema of conjunctiva, and acute keratoconjunctivitis. Slight lenticular opacities also occurred. Many developed liver damage, from which one died acutely. It is claimed that another died many years later from liver damage. On landing these men had been scrubbed and scrubbed, and chelating agents were used to try to remove the contamination. All their hair, including all their body hair, was shaved off before they were hospitalized; the hair and skinflakes were highly radioactive, and led to development of acute and long-term skin lesions (Figure 6).

Medical problems

Physicians must think clearly about what they could do in the event of a nuclear attack. In the post-attack period the first duty would be the care of the dying and the living. The awareness of the enormous numbers of dead or the grotesque destruction of the physical fabric of society, buildings, roads, hospitals, power and water supplies must not prevent the physician from thinking through what could be done.

When the consequences of a nuclear war are discussed, one must try to envisage what will be left after a nuclear attack to support survival, not for two weeks or two months but for years. One needs to calculate, and no doubt this has been done by every country preparing to defend itself in war, the optimum number of people to be allowed to survive. There must be enough in numbers and age distribution to work for recovery, but not so many as to swamp the restricted resources that may remain. Physicians need to know roughly how many survivors could be supported; but they are concerned not only about numbers but about the quality of survival. In the recovery period each survivor must be fit and well; otherwise, he is a large drain on the rest.

It is straightforward to predict the quality of survival following treatment of burns or breaks; but to predict the quality of survival following radiation injury, one would need to know something about the dose received. However, because the fallout will not be evenly distributed and radiation monitoring posts are likely to be several miles apart, the dose received by an individual will not be known even within a considerable error. Therefore one would have to guess the dose from the signs and symptoms.

In the post-attack situation it will be necessary to make a rough and rapid assessment of each case to channel resources to the group for which treatment is likely to result in survival, with lower priorities for the minimally or moderately exposed and for the presumably hopeless.

For estimating the severity of injury clinical and biological indicators are more important than dosage, but it is unlikely that laboratory facilities would be available for the biological indicators. Clinical signs and symptoms will have to be used. The presence or lack of nausea and vomiting tends to separate those who have been exposed to a high dose, possibly fatal, from those who have received a low dose, probably non-fatal. The vomiting due to radiation exposure is likely to begin between 20 minutes and 3 hours after exposure, and early onset suggests a high radiation dose. Individual episodes of vomiting may come on suddenly without preceding nausea. Diarrhea is another symptom, with very prompt and explosive diarrhea, particularly if bloody, likely to indicate a fatal dose.

The problem is that both vomiting and diarrhea can be caused by emotional stress, which will inevitably be predominant in the horrific aftermath of an attack. Redness of the conjunctiva may appear fairly promptly after a dose of 150 rads or more, and redness of the skin within a few hours after doses of 500 rads or more; but the doses at which either of these signs begin to manifest themselves are highly variable. If caused by beta-rays, they may seem more ominous than they are in reality, since the effect might be purely localized.

It is the dose to the bone marrow with which one needs to be primarily concerned, as this leads to the hematopoietic syndrome. This syndrome consists of alterations in blood formation due to damage to the stem cells in marrow and lymphatic tissues (Figure 7).[7] The changes may vary from mild to lethal. A decrease of lymphocytes occurs promptly, much of it taking place within the first 24 hours after exposure. The level of this early lymphopenia is one of the best indicators of severity of radiation injury.

In the dose range that produces the hematopoietic syndrome there is often an early rise in neutrophil granulocytes during the first 48 hours; but the numbers fall to fairly low levels at about day 10, followed by a transient abortive rise around day 15. The absence of an abortive rise is an unfavorable sign. Then there is a steady fall in the count, with a nadir at about day 30. If the patient survives, this is followed by spontaneous recovery beginning in the fifth week.

The time sequence of these changes is not altered much in relation to dose. With the highest dose the onset of pronounced granulocyte and platelet depression occurs earlier, but the time of recovery is only slightly delayed.

The problem is largely to keep the patients alive for about five weeks when marrow recovery will have begun. The two main mechanisms of death after acute radiation exposure are infection and hemorrhage, and in the crowded conditions of shelter life, infection will be a formidable problem. The two tend to develop simultaneously and to be synergistically and rapidly progressive. For example, areas of hemorrhage in the lungs provide foci favorable for bacterial growth, while infectious lesions in the intestinal wall may precipitate bleeding.

In this context it is salutary to recollect the procedure that is considered necessary when dealing with radiation exposure in peace time. Table 5 sets out a suggested plan for the control of infection following a radiation accident.[8] How could such a plan be even imagined in a postwar situation, when neither the necessary laboratory procedures and personnel nor the isolation facilities

Fig. 7. Typical chart of blood values in fairly severe hematopoietic syndrome

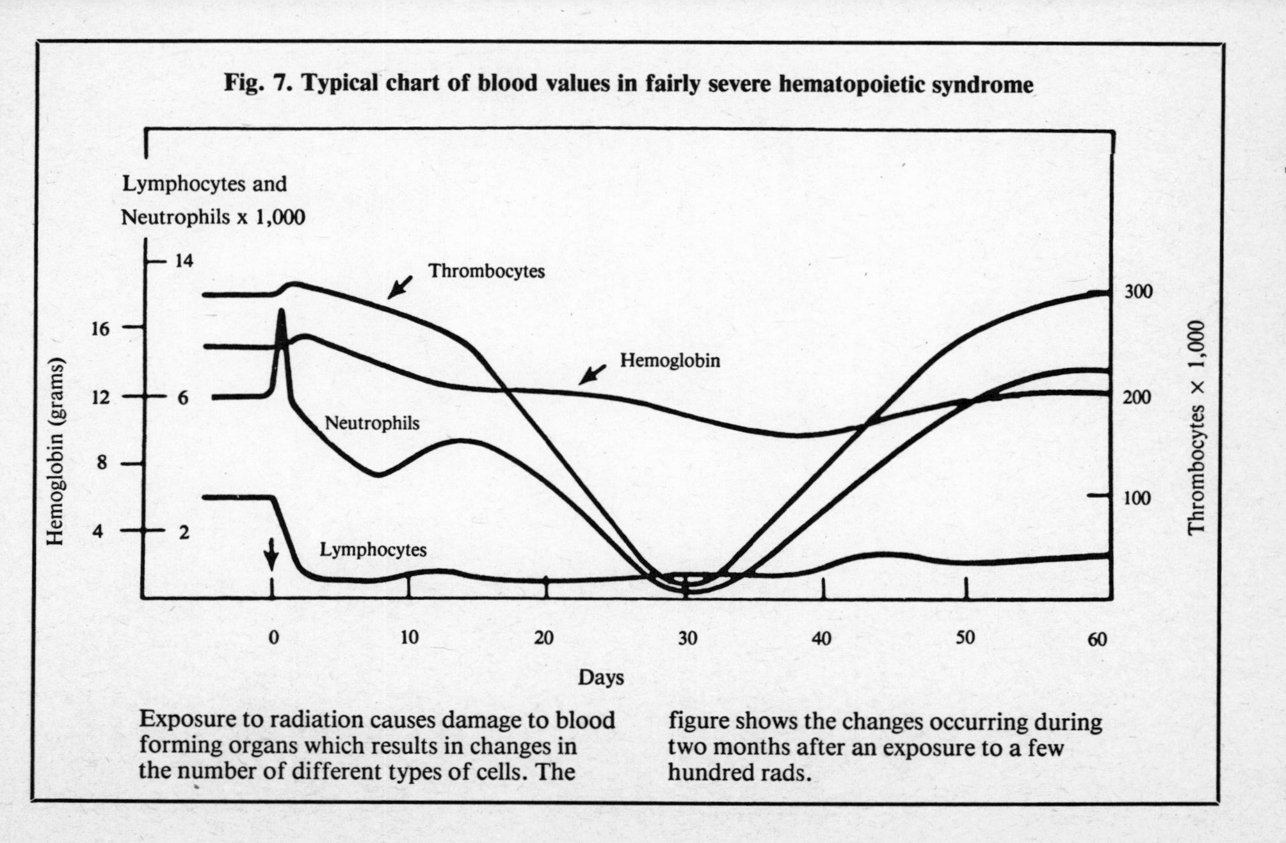

Exposure to radiation causes damage to blood forming organs which results in changes in the number of different types of cells. The figure shows the changes occurring during two months after an exposure to a few hundred rads.

Table 5

Suggested plan for control of infection

Immediately after diagnosis of exposure to 100 rad or more:

- Avoid hospitalizing patient except in sterile environment facility.
- Look for preexisting infections and obtain cultures of suspicious areas—consider especially carious teeth, gingivae, skin and vagina.
- Culture a clean-caught urine specimen.
- Culture stool specimen for identification of all organisms: run appropriate sensitivity tests for *Staph. aureaus* and gram-negative rods.
- Treat any infection that is discovered.
- Start oral nystatin to reduce *Candida* organisms.
- Do HLA typing of patient's family, especially siblings, to select HLA-matched leukocyte and platelet donors for later need.

If granulocyte count falls to less than 1,500 per cubic millimeter:

- Start oral antibiotics—vancomycin 500 mg liquid P.O. q. 4 hr.; gentamycin 200 mg liquid P.O. q. 4 hr; nystatin 1 million units liquid P.O. q. 4 hr., 4 million units as tablets P.O. q. 4 hr.
- Isolate patient in laminar flow room or life island.
- Daily antiseptic bath and shampoo with chlorhexidine gluconate.
- Trim finger and toenails carefully and scrub area daily.
- For female patients, daily Betadine douche and one nystatin vaginal tablet b.i.d.
- Culture nares, oropharynx, urine, stool, and skin of groins and axillae twice weekly.
- Culture blood if fever over 101°F.

If granulocyte count falls to less than 750 per cubic millimeter:

- In the presence of fever (101°F), or other signs of infection give antibiotics while awaiting results of new cultures, especially blood cultures. The regimen suggested is ticarcillin 5 gm q. 6 hr I.V.; gentamycin 1.25 mgm/kg q. 6 hr I.V.
- For severe infection not responding within 24 hrs, give supplemental white cells, and if platelet count is low, give platelets from preselected matched donors.
- When cultures are reported, modify antibiotic regime appropriately. Watch for toxicity from antibiotics, and reduce medications as soon as practicable.

When granulocyte count rises to over 1,000 per cubic millimeter and is clearly improving:

- Discontinue isolation, antiseptic baths, and antibiotics; continue nystatin for 3 additional days.

A typical plan for treating a potentially lethal hematologic syndrome.

would be available? It should also be remembered that procedures to prevent infection must begin on day 1 after exposure.

The use of antibiotics and fungicides will have to be carefully considered. Once infection has started in patients receiving doses of 200 rads or more, bone marrow failure is more rapid and severe. On the other hand, generalized use of fungicides and antibiotics, in populations confined in shelters, is likely to lead to selection of resistant infections, and to spread to other members of the group.

A judgment will need to be made between keeping antibiotics for those remaining in shelters—particularly those who might act as sources of infection (for example, the chronic bronchitic) even though they have not received a high radiation dose—or giving antibiotics to those who have already accumulated a dose of 200 rads, or who are expected to receive additional doses when going outside. These considerations are much more relevant to planning of postwar survival than arguments about the value of the LD-50 or other dosimetric parameters, particularly since it is unlikely that the dose received by a person will be known.

In order to create the impression that the situation can be handled, civil defense planners use numbers, and even unfounded mathematical formulas, to assess radiation injuries. For example, in the United Kingdom official civil defense plans consider 150 roentgens as acceptable exposure if the dose was received in two exposures separated by eight hours.[9] This is based on the assumption that there are two stages of response to radiation. The first stage, a relatively quick response to a dose received within about one day, awakens a dominant tolerance to exposure, making the body insensitive to an exposure up to about 150 roentgens. The second stage, a slow response, consists of the gradual replacement of cells killed by the previous radiation, by virtue of the division and growth of neighboring cells. This represents recovery from injury in the normal sense, assuming an estimated recovery rate of at least 10 roentgens per day.

In accordance with these plans, the wartime emergency dose will be 75 roentgens as a general rule, except that the persons engaged in vital tasks may undertake a second period of duty involving an additional 75 roentgens, provided there is a rest period of 8 hours between the two exposures. Persons who have remained in shelters for several days, and who have accumulated radiation doses, may undertake essential tasks, provided that the total exposure does not exceed 150 roentgens and is acquired over a period not exceeding 7 days.

For assessing the relative radiological states of people exposed for some time after a nuclear attack, the Operational Evaluation Dose is suggested. This is calculated from the following formula, applicable during the first 100 days after an attack:

$$OED = (X - 150 - 10t),$$

where X is the total accumulated dose expressed in roentgen units, and t is the time in days after the attack. According to this formula, if the calculated Operational Evaluation Dose has a negative or zero value, there will be no apparent radiation injury, and no physical or mental deterioration. If it is positive, then the Operational Evaluation Dose, and not the actual dose, would be used as a measure of the degree of injury.

The radiobiological bases for the assumption that the body can repair damage at a rate of 10 roentgens per day are hard to find and, indeed, are inconsistent with the description of radiation malaise given in *Nuclear Weapons.*[10]

> "On average radiation malaise will result from accumulated doses exceeding 150 r. Its symptoms are fatigue, nausea, indigestion, loss of appetite and, as the dose increases, there may also be vomiting, diarrhoea and the discharge of blood."

Clinical experience shows that although recovery from 150 rads is probable, repeated additional exposure to 10 rads per day is likely to result in severe bone marrow depression. Indeed, a dose of 10 rads per day over two weeks has been found in radiotherapy to suppress the bone marrow.

The persistence of excessive radiation levels is recognized in the U.K. Home Defence Circular:[11]

"7. . . . General life saving operations in areas of fallout might not be possible therefore until days or even weeks after a nuclear strike."

This, in effect, means that medical care will not be available in fallout areas to people who suffered injuries in the cramped conditions of life in shelters. So for days or weeks broken limbs will bleed and become infected; a trapped uninjured person will slowly die with acute radiation sickness—vomiting, diarrhea and hemorrhage—knowing that no one can come to him. It will not be a question of triage, as applied to conventional disasters. This will have been done by nature long before doctors and staff can enter the area.

When radiation fallout has decreased, however, it will be a question of triage of a different sort: selection of those cases for care who are known *not* to have received a high sub-lethal radiation dose, and who carry no contamination.

On what basis could physicians select for treatment those who had been in areas of heavy fallout?

It is not conceivable to use biological monitoring or whole body monitoring equipment to select those who should receive care because they had received little contamination. While others, who had been scrubbed and shaved, would have to be turned away if their internal contamination were too high, lest they contaminate not just the medical staff and equipment but their own families. The latter would need isolation facilities away from the family and community, and this would not be possible. One victim of an inhalation accident in peacetime can be successfully isolated in a caravan—many thousands cannot.

What would physicians do? Would it be on medical advice that these contaminated people would go back into the fallout areas to carry on rescue work? Would people agree to be measured for radioactive contamination if a high reading—which might be in error by 50 percent or more for gamma-radiation or by orders of magnitude for alpha- or beta-radiation—condemned them to death from radiation?

What criteria would physicians use to decide the radiation dose level at which an individual was not just expendable, but was an insupportable burden on limited resources? This is not a decision which can be abrogated to the health physicists or radiation monitors.

The ubiquitous nature of radioactive fallout, the unpredictability of its distribution, and its persistence, render useless any medical planning for dealing with the casualties of a nuclear attack. Underestimating the radiation problem for man himself, let alone for the food and resources on which his long-term survival depends, makes civil defense planning for the post-nuclear attack period a travesty of morality.

1. W.H. Langham, ed., *Radiobiological Factors in Manned Space Flight* (Washington, D.C.: National Academy of Sciences, 1967), p. 248.

2. C.A. Tobias and P. Todd, *Space Radiation Biology and Related Topics* (New York: Academic Press, 1974), p. 487.

3. S. Glasstone and P.J. Dolan, *The Effects of Nuclear Weapons* (Washington, D.C.: U.S. Department of Defense, 1977), p. 436.

4. G. Johnson, "Paradise Lost," *Bulletin of the Atomic Scientists,* 36: 10 (1980), 24.

5. Johnson, "Paradise Lost."

6. T. Kumatori, T. Ishihara, K. Hirashima, H. Sugiyama, S. Ishii, and K. Miyoshi "Follow-up Studies Over a 25-year Period on the Japanese Fishermen Exposed to Radioactive Fallout in 1954," in *The Medical Basis for Radiation Accident Preparedness,* edited by K.F. Hübner and S.A. Fry (North Holland: Elsevier, 1980), p. 33.

7. G.A. Andrews, "Medical Management of Accidental Total-Body Irradiation," in *The Medical Basis for Radiation Accident Preparedness,* edited by K.F. Hübner and S.A. Fry (North Holland: Elsevier, 1980), p. 297.

8. Andrews, "Medical Management."

9. *Nuclear Weapons,* ISBN 011 340557X (London: Her Majesty's Stationery Office, 1980).

10. *Nuclear Weapons.*

11. *Home Defence Circular* (77)1 (London: D.H.H.S., 1977).

11 Occurrence of cancer in atomic bomb survivors

STUART C. FINCH

The atomic bomb survivors of Hiroshima and Nagasaki constitute the largest population of persons known to have been exposed to large amounts of ionizing radiation. They have been under intensive and careful medical observation by a dedicated team of Japanese and American scientists of the Atomic Bomb Casualty Commission and the Radiation Effects Research Foundation for over 30 years. The most reliable direct information of the adverse late medical effects for man has been provided through these studies.

It now is clear that the most important late medical effect of atomic bomb radiation exposure is the increased occurrence of cancer. A number of other effects of much smaller magnitude, of uncertain consequence, or of uncertain relationship to radiation exposure, also have been observed. Several reviews in recent years have summarized in some detail the results of all studies from the Casualty Commission and the Research Foundation.[1-4]

* * *

On August 6, 1945, a uranium-235 bomb was exploded over Hiroshima at about 570 meters above ground. The bomb had an explosive yield of about 12.5 kilotons (thousands of tons of TNT equivalent). Three days later, a plutonium-239 bomb was exploded about 510 meters above Nagasaki, with an explosive yield of about 22 kilotons. Both blasts lasted from 0.5 to 1 second and produced heat intensities of 3,000 to 4,000 degrees Centigrade at ground zero. Blast pressures were estimated to be in the range of 4.5 to 8 tons per square meter.

Virtually all of the ionizing radiation received by the atomic bomb survivors occurred almost instantly as the bombs were detonated. There was relatively little fallout or ground contamination since both bombs were exploded at a considerable distance above ground. Most of the radiation released from the bombs was gamma, but the structure of the Hiroshima bomb was such that a significant component of the radiation released was neutron in type. Although some controversy currently exists regarding the neutron-gamma ratio of the radiation actually received by Hiroshima survivors, it is unlikely that the well-established total dose estimates for each survivor will be changed significantly.[5,6] Also, the proposed changes in Hiroshima dose estimates would have little or no effect on the dose-response relationships which have been established in both cities.[6]

A team of Japanese and American scientists in the fall of 1945 estimated that approximately 64,000 civilians were either killed immediately or died within about two months in Hiroshima.[7] No reliable data are available on losses among the military, but it is known that a large military population was stationed near the hypocenter. The corresponding estimate for civilian deaths at Nagasaki was about 40,000. Mortality estimates by others have been considerably higher for both cities. Deaths were about equally due to blast, burn, or radiation exposure.

The Atomic Bomb Casualty Commission began its work in Japan in 1947. Funded by the U.S. Atomic Energy Commission, it was under the direction of the National Academy of Sciences-National Research Council, for the purpose of long-term study of late radiation effects in both cities. The Japanese National Institute of Health formally joined the studies in 1948. The Commission continued as a joint undertaking of the National Academy of Sciences and Japanese National Institute of Health until April of 1975 at which time it was dissolved, to be replaced by the Radiation Effects Research Foundation, an independent research organization funded equally by Japan and the United States. This successful bi-national research project is unique in the history of cooperative international human research, and is a testimonial to the recognized importance of understanding all late radiation effects in man.

The early scientific efforts by the Atomic Bomb Casualty Commission consisted of a series of ad hoc studies to answer specific questions. Several clinical samples were established for the purpose of detecting possible somatic effects of radiation exposure. A program for procurement of autopsies was started and an intensive search for possible genetic effects was conducted. Opthalmologic observations were made, the leukemia situation was explored and the children who were exposed while *in utero* were checked for evidence of physical disability.

In those early days, different population samples for each study frequently were constructed. The results were productive, but some questions were raised in the mid-1950s concerning the possibility of undertaking a more systematic approach to the detection of late radiation effects.

In 1955, a committee of National Academy of Sciences-National Research Council consultants reviewed the Casualty Commission research design and recommended the establishment of fixed and integrated population samples for long-term population studies—the Unified Program.[8] This resulted in the consolidation of a number of study populations into a single population sample which became the focus of studies of duration of life, mortality from specific disease, and the occurrence of medical effects during life. *In utero* and F_1 (first generation offspring) samples also were established for long-term mortality and clinical studies. Each study population included individuals with different degrees of radiation exposure, as well as unexposed matched controls, against which radiation effects could be measured. Minor modifications in the size and composition of the samples have occurred over the years, but they have essentially remained intact.

The Extended Life Span Study, the largest of the Unified Program study populations, included about 110,000 survivors and their controls who were alive at the time of the 1950 A-bomb Survivors Survey. Deaths within this cohort have been ascertained with considerable accuracy by periodic searches of the Family Registry (Koseki) which makes it possible to secure information on the deaths of survivors, even though there is considerable population movement.

The base population for the Pathology Study consists of a subset of persons in the Life Span Study who reside in the Hiroshima-Nagasaki area and are candidates for postmortem studies at the time of death. Autopsy rates as high as 45 percent in the early 1960s have provided information of great value in confirmation of death certificate diagnoses and the histologic description of radiation-induced tumors. In recent years, the tumor and tissue registries of both cities have largely replaced autopsy studies in the evaluation of radiation tumor induction.

The Adult Health Study is a biennial health examination program for a Life Span Study subset population of about 20,000 persons. This sample includes most of all persons known to have had high radiation dose exposure, the *In Utero* Clinical Sample, and appropriate non-exposed controls. The purpose of the Adult Health Study is to determine disease morbidity and to search for clinical evidence of tissue change due to radiation exposure in atomic bomb survivors. Remarkably, about 80 percent of those who are available for examination continue to return for regular re-examinations.

Decisive to the construction of dose-response relationships is the amount of radiation exposure.[9,10,11] The earliest studies at the Atomic Bomb Casualty Commission used distance from the hypocenter and the severity of certain acute radiation symptoms as measures of radiation exposure, but in the mid-1950s, methods for estimating radiation dose received under different shielding conditions were developed.[9] Further refinements resulted in the 1965 system of dose estimates which currently are used.[10] Radiation dose estimates now have been completed for all but about 3 percent of the persons in the Life Span Study sample. Dose estimates are expressed in rad, as tissue kerma (kinetic energy released in materials) in air. It is believed that total radiation dose estimates for persons in the Radiation Effects Research Foundation studies are accurate to within plus or minus 10 to 15 percent. These estimates represent great improvement over distance from the hypocenter as a measure of "air dose." Although modest revisions in dose estimates may occur in the future, it is unlikely that these changes will have any appreciable effects on current dose-response results.

The task of detecting late effects of exposure to ionizing radiation is difficult for many reasons. Perhaps the most important is the absence of disease which is uniquely the result of radiation exposure. Also, in most cases, radiation exposure is just one of the competing factors in carcinogenesis. The problem of determining those effeccts that are due to other types of environmental exposure becomes quite formidable. Thus, in all studies of this type, the determination of late radiation effects is made by comparing the experience of the survivors with the experience of an appropriate control group. The detection of a significant regression coefficient for a dose-response relationship is the single most useful technique for the establishment of a radiation-related effect.

Mortality ascertainment for the Life Span Study began in 1950 and does not cover the population that died during the five-year period before the sample was selected. Most of those deaths were due to acute thermal, blast and radiation effects. There is some evidence that the other deaths which occurred prior to sample selection were not excessive and have not been a serious source of bias.[12]

A number of definite and borderline radiation-induced somatic effects have been observed in Hiroshima and Nagasaki survivors (Tables 1 and 2). Results to date of other important studies have failed to demonstrate any radiation effect (Table 3). The information here, however, will be confined to radiation-induced cancer.

An increase in leukemia incidence was first noted about three years after exposure to the A-bombs and reached a peak around 1952 to 1953.[13] All forms of leukemia, with the exception of chronic lymphocytic leukemia, appear to have increased in the exposed persons; but there are complex differences among the types of leukemia in relationship to age at time of bomb, city of exposure, and duration of the latent period following exposure.

The leukemia incidence has been high among those who were heavily exposed in every age group. The relative risk for those exposed to doses of 100 rad or more, however, was the highest for those who were under 10 or over 50 years of age at the time.[13,14] The ratio of 20:1 for these age groups may be compared to ratios of about 14:1 for the other age groups. Age also had an effect on the type of leukemia which developed. For those who were 30 years or older at the time of exposure the early excess was more pronounced for chronic than for acute leukemia. For those who were under 15 years, most of the excess was of the acute lymphocytic type, but there also was a significant increase in chronic granulocytic leukemia in this group.

Table 1

Late effects for which a definite relationship has been established with atomic bomb exposure

- Increased occurrence of cataracts and other lens changes
- Increased incidence rates for leukemia, multiple myeloma, and cancers of the thyroid, breast, lung and stomach
- Increased frequency of blood lymphocyte chromosome aberrations
- Increased small head size and mental retardation following *in utero* exposure
- Impaired growth and development following *in utero* or childhood exposure
- Increased tumor incidence following exposure during early life

A clear relationship between the incidence of leukemia and radiation dose has been present for both cities, but the effect was more clear-cut in Hiroshima than in Nagasaki. The lowest doses with a demonstrable leukemogenic effect appear to be in the 20-to-30 rad range in Hiroshima.[15] The difference between the Hiroshima and Nagasaki experiences, especially in the lower dose ranges, has been attributed in the past to a relatively large neutron component of the ionizing radiation in Hiroshima.[13] The relative-

Table 2

Late effects for which there are borderline or suggestive relationships with atomic bomb exposure

- Tumors of the esophagus, colon, salivary glands and urinary tract organs
- Malignant lymphoma
- Myelofibrosis
- Certain aspects of tumoral and cell-mediated immunity
- Some age-related tissue changes

Table 3

Late effects for which no significant relationship with atomic bomb exposure has been shown

- F_1 – increase in mortality, birth defects, leukemia, or alterations of growth and development
- Infertility
- Disease other than neoplasms
- Increased susceptibility to illness
- Life shortening, exclusive of neoplasia

ly small size of the Nagasaki population at risk, however, may be more responsible for the unclear relationships at low dose than was the quality of the radiation absorbed. The question of whether a linear relationship exists between radiation dose and leukemia incidence for radiation doses below 20 rad continues to remain somewhat controversial.

Age at the time of the bomb was an important factor in determining the duration of the latent period between radiation exposure and the occurrence of acute leukemia.[13] Those individuals who were less than 15 years old had a relatively short latent period for acute leukemia, virtually all of which occurred in the early 1950s. In contrast, those who were 45 years of age and over had a considerably longer latent period for acute leukemia with continuation of the leukemogenic effect through the 1960s and into the early 1970s. For chronic granulocytic leukemia, the highest rates have been observed in the heavily exposed who were young at the time of exposure, but there is little evidence of an age differential in the latent period of the occurrence of this form of the disease. The peak incidence for chronic granulocytic leukemia occurred in the earlier years following exposure regardless of age at the time of the bomb.

There is no evidence of a radiation-induced leukemogenic effect in persons who were *in utero*, despite the fact that the highest leukemia rates have occurred in individuals who were young at the time of exposure.[16] Furthermore, no excess in leukemia incidence has been observed up to the present time among the children born of exposed parents.[17]

Leukemia rates have declined steadily in both cities since 1952. In Nagasaki, the rate in the exposed survivors has not exceeded that of the control population since the early 1970s; in Hiroshima, there is evidence of continuation of a slightly increased leukemia rate in the exposed.[14] In contrast to the decline in leukemia rates, the death rates from solid tumors of various sites in exposed individuals have increased considerably in recent years.[14,18] The ac-

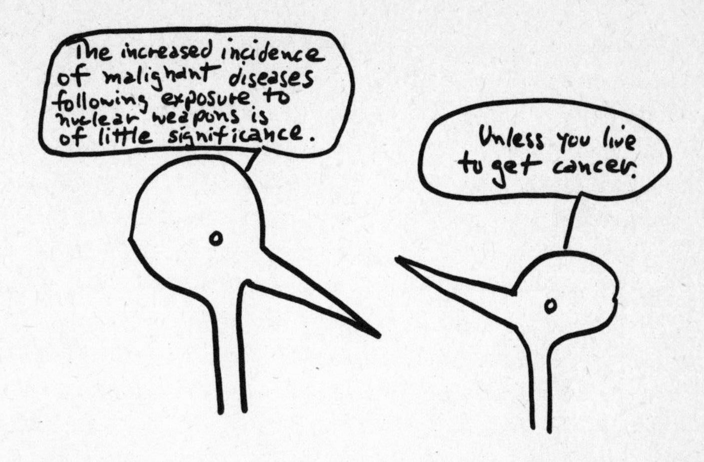

tual total number of radiation-induced solid tumors now is in excess of the number of radiation-induced leukemias in the exposed populations of the two cities.

In the late 1950's and early 1960's, clinical studies at the Atomic Bomb Casualty Commission showed that thyroid tumors were occurring more frequently among survivors, especially in women who had been exposed to high radiation doses, than in those who had received little or no radiation.[19] During the period 1958 to 1971 the relative risk for thyroid cancer for persons exposed to 100 rad or more was about 2.5 times that of the controls.[20] No relationship has been shown between either age at the time of the bomb or dose and time of onset. Most of the radiation-related malignant tumors have been clinically evident papillary carcinomas which rarely have metastasized or caused disability.[20,21] A modest radiation exposure relationship also has been established for small occult papillary carcinomas of the thyroid.[20]

In 1968, the first report from the Casualty Commission appeared linking A-bomb exposure to cancer of the female breast.[22] A subsequent more intensive study showed that the effect started in the mid 1950s for women in the higher exposure group. The dose-response relationships in both cities appear to be linear and are of similar magnitude. During the period 1950 to 1974 the age adjusted relative risk for women exposed to 100 rad or more was about 3.3 in comparison to the controls.[23] The highest relative risk has occurred in women who were ages 10 to 19 when exposed, and who received at least 100 rad.[24] The incidence of breast cancer among women has increased in heavily exposed women in all other age groups with the exception of ages 0 to 10 and 40 to 50. It is not possible at this time to explain these differences in age susceptibility to radiation-induced breast cancer, but the information does suggest that hormonal factors may be important. No unusual or specific types of breast cancer have been associated with previous radiation exposure.[23]

A carcinogenic effect of ionizing radiation on the lung was shown in 1965 and again in 1968 for men who were 35 to 40 years of age or older at the time of the bomb.[25,26] The effect began about 1955 and was very definite by 1960 in Hiroshima, but was much less striking in Nagasaki. The most recent report on lung cancer deaths indicates that the relative risk of lung cancer for those exposed to 100 rad or more is about 1.8 times what was expected,[14,27] and it appears to be independent of smoking.[28] Small cell anaplastic carcinomas are increased with radiation dose. Also, there was a slight increase in risk of epidermoid carcinoma and bronchogenic adenocarcinoma among exposed individuals, but these increases were not statistically significant.[27]

Mortality figures for stomach cancer demonstrate a moderately increased risk for persons exposed to 100 rad or more.[14] Most of the excess stomach cancer has occurred in Hiroshima; an increased risk in Nagasaki is evident only at high exposure levels.[29] Some very recent studies at the Radiation Effects Research Foundation indicate that the relative risk for multiple myeloma in persons exposed to 100 rad or more is significantly increased in both cities, especially for persons who were aged 20 to 40 at the time of the bomb.[30,31] The excess risk for multiple myeloma in this high-

Table 4

Cancer risk estimates for atomic bomb survivors, ages, sexes, and cities combined

Malignancy	*Time interval*	*Type of data*	*Relative risk*[a] *100 rads or more*	*Absolute risk*[a] *excess deaths per million person-year-rad*	*Reference*
Leukemia	1950 to 1974	Mortality	10.5 (8.3-13.3)	1.92 (1.76-2.07)	14
Breast (female)	1950 to 1974	Incidence	3.3 (2.6-4.1)	0.36 (0.10-0.62)	14,24
Thyroid	1950 to 1971	Incidence	2.5 (1.6-4.0)	—	20
Urinary tract	1950 to 1974	Mortality	2.2 (1.3-3.1)	0.13 (0.02-0.25)	14
Lymphoma	1950 to 1974	Mortality	1.8 (1.1-2.7)	0.18 (0.06-0.29)	14
Lung	1950 to 1974	Mortality	1.8 (1.4-2.2)	0.35 (0.12-0.58)	14
Esophagus	1950 to 1974	Mortality	1.6 (1.0-2.2)	0.19 (0.04-0.33)	14
Stomach	1950 to 1974	Mortality	1.2 (1.0-1.3)	0.64 (0.17-1.12)	14

[a] *Relative risk* figures are for times the normal risk for persons receiving 100 or more rads in the period indicated; 80 percent confidence intervals for relative risk. For example, the average relative risk of leukemia is 10.5 times the normal risk for a person receiving 100 or more rads from 1950 to 1974.

Absolute risk figures are for times the normal risk for persons receiving one rad in a year; 90 percent confidence intervals for absolute risk.

dose group first became apparent about 20 years following exposure.

There are several borderline or suggestive radiation-related tumors that currently are under intensive investigation. An increased risk for malignant salivary gland tumors has been reported for persons exposed to 300 rad or more in Hiroshima only.[32] Current mortality data suggest a modest radiation effect for lymphoma and cancers of the esophagus, colon and urinary bladder.[14]

No significant relationship has been established to date between radiation exposure in atomic bomb survivors and tumors of the brain, pancreas, gallbladder, bile duct, bone, skin, rectum, uterus or prostate. Children exposed to 100 rad or more of radiation have developed an excessively large number of malignant tumors.[33] The actual number of cancer deaths in this group, however, is relatively small and there have been few additions in recent years. Increased mortality from solid tumors has not been demonstrated, however, in either the *in utero* exposed or the children born to exposed parents.[34]

The joint studies of the Commission-Foundation make clear the effect of ionizing radiation for cancer induction in man. (Table 4). Although the magnitude of the carcinogenic effect is not sufficient to influence significantly the overall survival of the exposed population, the data accumulated thus far constitute the largest single body of information known to man on the risk of radiation induced cancer in man. Such information has been of great value in the establishment of radiation safety standards for man and, one hopes, as a deterrent to the future use of atomic energy as an instrument of war.

1. S. C. Finch, "The Study of Atomic Bomb Survivors in Japan," *The American Journal of Medicine* 66 (1979), pp. 899-901.

2. G. W. Beebe, "Reflections on the Work of the Atomic Bomb Casualty Commission in Japan," *Epidemiologic Reviews* 1 (1979), pp. 184-210.

3. S. C. Finch and I. M. Moriyama, "The Delayed Effects of Radiation Exposure among Atomic Bomb Survivors, Hiroshima and Nagasaki, 1945-79: A

Brief Summary," *Journal of the Hiroshima Medical Association* 33 (1980), pp. 83-94.

4. S. Okada and others, "A Review of 30 Years Study of Hiroshima & Nagasaki Bomb Survivors," *Journal of Radiation Research* 16, supplement (Tokyo: 1975), pp. 1-164.

5. E. Marshall, "New A-Bomb Studies Alter Radiation Estimates," *Science* 212 (1981), pp. 900-903.

6. S. Jablon, W.E. Loewe, E. Mendelsohn; R. L. Dobson, T. Straume; D.C. Kaul, "Radiation Estimates" (letters to the editor), *Science* 213 (1981), pp. 6-8.

7. A.W. Oughterson, S. Warren, "Medical Effects of the Atomic Bomb in Japan," *National Nuclear Energy Series*, Div. VIII (New York: McGraw-Hill, 1956).

8. T. Francis, S. Jablon, F.E. Moore, "Report of Ad Hoc Committee for Appraisal of ABCC Program" (1955), ABCC TR, pp. 33-59.

9. E. T. Arakawa, "Radiation Dosimetry in Hiroshima and Nagasaki Atomic Bomb Survivors," *New England Journal of Medicine* 263 (1960), pp. 488-93.

10. R.C. Milton, T. Shohoji, "Tentative 1965 Radiation Dose Estimation for Atomic Bomb Survivors, Hiroshima-Nagasaki," ABCC TR, pp. 1-68.

11. J. A. Auxier, "Ichiban, Radiation Dosimetry for the Survivors of the Bombings of Hiroshima and Nagasaki," *ERDA Critical Review Series* (1977).

12. K. Tachikawa, H. Kato, "Mortality among Atomic Bomb Survivors, October 1945-September 1964," based on 1946 Hiroshima City Casualty Survey, ABCC TR, pp. 6-69.

13. M. Ichimaru, T. Ishimaru, "Review of 30 Years Study of Hiroshima and Nagasaki Atomb Bomb Survivors, II. Biological Effects, D. Leukemia and Related Disorders," *Journal of Radiation Research* 16, supplement (Tokyo: 1975), pp. 89-96.

14. G.W. Beebe, H. Kato, C.E. Land, "Studies of the Mortality of A-Bomb Survivors," 6. Mortality and Radiation Dose, 1950-1978," *Radiation Research* 75 (1978), pp. 138-201.

15. T. Ishimaru, T. Hoshino, M. Ichimaru, H. Okada, T. Tomiyasu, T. Tsuchimoto, T. Yamamoto, "Leukemia in Atomic Bomb Survivors, Hiroshima and Nagasaki, 1 October 1950-30 September 1966," *Radiation Research* 45 (1971), pp. 216-33.

16. S. Jablon, H. Kato, "Childhood Cancer in Relation to Prenatal Exposure to A-Bomb Radiation," *Lancet* 2 (1970), pp. 1,000-3.

17. T. Hoshino, H. Kato, S.C. Finch, Z. Hruber, "Leukemia in Offspring of Atomic Bomb Survivors," *Blood* 30 (1967), pp. 719-30.

18. G.W. Beebe, H. Kato, "Review of 30 Years Study of Hiroshima and Nagasaki Atomic Bomb Survivors, II. Biological Effects, E. Cancers other than Leukemia," *Journal of Radiation Research* 16, supplement (Tokyo: 1975), pp. 97-107.

19. E.L. Socolow, A. Hashizumi, S. Neriishi, R. Niitani, "Thyroid Carcinoma in Man after Exposure to Ionizing Radiation: A Summary of the Findings in

Hiroshima and Nagasaki," *New England Journal of Medicine* 268 (1963), pp. 406-10.

20. L.N. Parker, J.L. Belski, T. Yamamoto, S. Kawamoto, R.J. Keehn, "Thyroid Carcinoma after Exposure to Atomic Radiation: A Continuing Survey of Fixed Population, Hiroshima and Nagasaki, 1968-1971," *Annals of Internal Medicine* 80 (1974), pp. 600-4.

21. R.J. Sampson, C.R. Key, C.R. Buncher, S. Iijima, "Thyroid Carcinoma in Hiroshima and Nagasaki, 1. Prevalence of Thyroid Carcinoma at Autopsy," *Journal of the American Medical Association* 209 (1969), pp. 65-70.

22. C.K. Wanebo, K.G. Johnson, K. Sato, T.W. Thorslund, "Breast Cancer after Exposure to the Atomic Bombings of Hiroshima and Nagasaki," *New England Journal of Medicine* 279 (1968), pp. 667-71.

23. D.H. McGregor, C.E. Land, K. Choi, S. Tokuoka, P.I. Liu, T. Wakabayashi, G.W. Beebe, "Breast Cancer Incidence among Atomic Bomb Survivors, Hiroshima and Nagasaki, 1950-69," *Journal of the National Cancer Institute* 59 (1977), pp. 799-811.

24. M. Tokunaga, J.E. Norman, Jr., M. Asano, S. Tokuoka, H. Ezaki, I. Nishimori, Y. Tsuji, "Malignant Breast Tumors among Atomic Bomb Survivors, Hiroshima and Nagasaki, 1950-1974," *Journal of the National Cancer Institute* 62 (1979), pp. 1374-59 (RERF TR 17-77).

25. A. Ciocco, "JHNIH-ABCC Life Span Study and ABCC-JNIH Adult Health Study: Mortality 1950-64 and Disease Survivorship 1958-64 among Sample Members Age 50 Years or Older, 1 October 1950," ABCC TR, pp. 18-65.

26. C.K. Wanebo, K.G. Johnson, K. Sato, T.W. Thorslund, "Lung Cancer Following Atomic Radiation," *American Review of Respiratory Diseases* 8 (1968), pp. 778-87.

27. R.W. Chihak, T. Ishimaru, A. Steer, A. Yamada, "Lung Cancer at Autopsy in A-Bomb Survivors and Controls, Hiroshima and Nagasaki, 1961-1970; 1. Autopsy Findings and Relation to Radiation," *Cancer* 33 (1974), pp. 1580-8.

28. T. Ishimaru, R.W. Chihak, C.E. Land, A. Steer, A. Yamada, "Lung Cancer at Autopsy in A-Bomb Survivors and Controls, Hiroshima and Nagasaki, 1961-1970, 2. Smoking, Occupation and A-Bomb Exposure," *Cancer* 36 (1975), pp. 1723-8.

29. K. Nakamura, "Stomach Cancer in Atomic-Bomb Survivors" (letters to the editor), *Lancet* 2 (1977), pp. 866-7.

30. M. Ichimaru, T. Ishimaru, M. Mikami and others, "Incidence of Multiple Myeloma among Atomic Bomb Survivors by Radiation Dose, Hiroshima and Nagasaki, 1950-76" (abstract), *Acta Haematol* 41 (Japan: April 1978), p. 219.

31. J. Cuzick, "Radiation Induced Myelomatosis," *New England Journal of Medicine* 304 (Jan. 22, 1981), pp.204-10.

32. J.L. Belski, K. Tachikawa, R.W. Cihak, T. Yamamoto, "Salivary Gland Tumors in Atomic Bomb Survivors, Hiroshima-Nagasaki, 1957 to 1970," *Journal of the American Medical Association* 219 (1972), pp. 864-8.

33. S. Jablon, K. Tachikawa, J.L. Belski, A. Steer, "Cancer in Japanese Exposed as Children to Atomic Bombs, *Lancet* 1 (1971), pp. 927-32.

34. H. Kato, W.J. Schull, J.V. Neel, "Survival in Children of Parents Exposed to Atomic Bomb" (A Cohort-Type Study) *American Journal of Genetics* 18 (1966), pp. 339-73.

IV PROGNOSIS

"The medical facilities of the nation would choke totally on even a fraction of the resulting burn casualties alone."

—Herbert L. Abrams

Our relationship to nuclear weapons is unhealthy.
Let us not get too abstract.

12 The clinical picture

HOWARD H. HIATT

When I was invited to introduce a symposium entitled The Medical Consequences of Nuclear War, I was somewhat skeptical about the theme of the conference and even more skeptical about my suitability for participation. On both counts my skepticism was quickly put to rest. Increasing talk by public figures about winning or even surviving a nuclear war, I concluded, must reflect their unawareness of medical realities: nuclear war would inevitably lead to death, disease and suffering of epidemic proportions, and without effective medical interventions. Those realities, in turn, inevitably lead to the same conclusion we have reached for such contemporary epidemics as those of lung cancer and heart disease: prevention is critical for effective control.

What can be said about the epidemics that would result from the use of nuclear weapons? Two sources of information are available: descriptions of the medical effects of the Hiroshima and Nagasaki bombs and theoretical projections of the medical effects of bombing American (or Soviet) cities toward which Soviet (or American) nuclear weapons are now aimed.

The Hiroshima bomb, the explosive power of which was equivalent to 13,000 tons of TNT, is estimated to have killed 70,000 of the population of 245,000 and destroyed two-thirds of the 90,000 buildings within the city limits. Perhaps even more devastating than the statistics are the descriptions of individual victims. Consider this picture presented by John Hersey in his book, *Hiroshima:*

"There were about 20 men . . . all in exactly the same nightmarish state: their faces were wholly burned, their eye sockets were hollow, the fluid from their melted eyes had run down their cheeks. . . . their mouths were mere swollen, pus-covered wounds, which they could not bear to stretch enough to admit the spout of a teapot. . . ."

In 1962, several articles in the *New England Journal of Medicine* used a study prepared three years earlier by the Joint Congressional Committee on Atomic Energy to examine the effects of a hypothetical nuclear attack on Boston.

Boston's disaster begins with a 20-million ton bomb exploding at ground level in the city's downtown area and excavating a crater one-half mile in diameter. The circle within which even the most heavily reinforced concrete structures do not survive has a radius of four miles and includes most of the hospitals and medical personnel in the area. As far as 15 miles from the blast, all frame buildings are damaged beyond repair.

Detonation of the bomb releases a tremendous amount of thermal energy. For up to 40 miles away, retinal burns from looking at the fireball cause blindness. More than 20 miles from the center, the firestorm, fueled by igniting houses, foliage, and oil and gasoline storage tanks, accounts for many deaths and injuries and increases the already catastrophic physical damage caused by the blast.

The blast and firestorm cause 2.2 million fatalities; survivors are badly burned, blinded, and otherwise seriously wounded. Many are disoriented. The need is great for medical care, food, water, shelter and clothing, but all are gravely inadequate. These are the short-term effects; the problem of radiation sickness will grow in the period ahead.

How would modern medicine deal with the casualties of a nuclear attack? Hersey described the problems presented to Hiroshima's medical care system and its capabilities and response:

"Of 150 doctors in the city, 65 were already dead and most of the rest were wounded. Of 1,780 nurses, 1,654 were dead or too badly hurt to work. In the biggest hospital, that of the Red Cross, only six doctors out of 30 were able to function, and only 10 nurses out of more than 200. . . . At least 10,000 of the [city's] wounded made their way to the [Red Cross Hospital] which was altogether unequal to such a trampling. . . ."

In the aftermath of a nuclear attack on Boston, what are the prospects for medical care? Using as their base a figure of 6,560 physicians in the Boston metropolitan area at the time of attack, the 1959 and 1962 studies project that almost 5,000 will be killed immediately or fatally injured, and that only 900 will be in condition to render post-attack medical care. The ratio of injured persons to physicians is thus in excess of 1,700 to one. If a physician spends an average of only 15 minutes with each injured person and works 16 hours each day, the studies project, it will take from 16 to 26 days for each casualty to be seen once.

Thus, it is unrealistic to suggest a meaningful medical response to the overwhelming health problems that would follow a nuclear attack. Similarly, only the most limited medical measures can be visualized to deal with the burden of cancer and genetic defects that would afflict survivors and future generations. With respect to the practicality of temporary evacuation, radioactivity would make the blast area uninhabitable for months.

At present more than 50,000 nuclear weapons are deployed and ready. Most dwarf in destructive power the bomb used against Hiroshima. Sufficient nuclear bombs exist outside the United States to subject every major American city repeatedly to the destruction described for Boston.

Preparing for the symposium was a stressful experience for me. What would be the purpose, I wondered initially, in describing such almost unthinkable conditions? But the conditions are not unthinkable; rather they are infrequently thought about, much less discussed. Among the painful results of the silence are the continuing proliferation of nuclear weapons and the failure to reject out-of-hand nuclear war as a "viable option" in the management of world problems.

If we examine the consequences of nuclear war in medical terms, we must pay heed to the inescapable lesson of contemporary medicine: where treatment of a given disease is ineffective *or* where costs are insupportable, attention must be given to prevention. Both conditions apply to the effects of nuclear war—treatment programs would be virtually useless and the costs would be staggering. Can more compelling arguments be marshalled for a preventive strategy?

Prevention of any disease requires an effective prescription. It is clear that such a prescription must not only prevent nuclear war but also safeguard our security. The knowledge and credentials of physicians do not give us any special competence in the discussion of security issues. However, if our political and military leaders have based strategic planning on mistaken assumptions concerning the medical aspects of a nuclear war, we do have a responsibility. We must inform them and the American people of the full-blown clinical picture that would follow a nuclear attack and the impotence of the medical community to offer a meaningful response. To remain silent is to risk betraying ourselves and our country.

13 Illusion of survival

H. JACK GEIGER

To attempt to measure and describe the consequences of a thermonuclear attack on a major American city is to confront a paradox.

On the one hand, the nature and magnitude of the effects of hypothetical—but eminently possible—nuclear attacks are entirely specifiable. The calculations are straightforward and only moderately complex. Indeed, over the past two decades, these consequences have been described in exquisite detail in hundreds of scientific journals (including *The Bulletin of the Atomic Scientists*), books and government publications.[1]

On the other hand, despite the specificity, these effects—the numbers of killed and injured, the destruction of the physical environment, the damage to the ecosphere—are unfathomable. In short, it is almost impossible fully to grasp the reality they represent, the implications they carry.

This is not merely because the numbers are so large as to be incomprehensible: close to 10 *million* people killed or seriously injured, for example, in consequence of a single 20-megaton explosion and the resulting firestorm on the New York metropolitan area. The difficulty occurs primarily because we are attempting to describe and understand an event that is without human precedent.

Hiroshima and Nagasaki do not serve as precedents for any probable nuclear war scenario. The weapons used on those cities approximated 13 kilotons of explosive force each. At one megaton—a small weapon by contemporary standards—we are trying to imagine 70 simultaneous Hiroshima explosions. At 20 megatons we are trying to imagine 1,400 Hiroshima bombs detonated at the same moment in the same place.

Hiroshima and Nagasaki were single events, with effects decaying over time; today we are faced with the possibility of multiple events—a thermonuclear explosion at 10 a.m. and another at 4 p.m. At the time of Hiroshima, there was one nuclear power and the world's total arsenal comprised two or three weapons; today there are at least six nuclear powers and the total arsenal is—conservatively—in excess of 50,000 warheads.

But most important, Hiroshima and Nagasaki were isolated, limited disasters. They could, in time, be saved and reconstructed with help from outside. Always, we think of an "outside"; this is our intuitive model of disasters, for our historical experience is confined to single-event phenomena of limited range, duration and effect—hurricanes, earthquakes, World War II bombings, even Hiroshima and Nagasaki—in which both short-term and longer-term relief efforts could be mounted.

In any full-scale contemporary nuclear exchange, however, *there will be no "outside" that we can rely upon.* We cannot safely assume that there will be unaffected major areas within reach of targeted cities that will have resources that can be mobilized effectively to help the stricken targets, or that are likely to regard even making the effort as a rational enterprise. In a population-targeted attack, every major population center may be effectively destroyed.

In attempting to comprehend the consequences it is useful to consider the case of a single weapon and a single city. One specifies the megatonnage, the nature of the attack (air burst or ground burst, single or multiple strike), the time of year, the day of the week (workday or weekend), the time of day, the atmospheric conditions (clear or cloudy, raining or dry), and the wind patterns. The magnitudes of blast (in pounds per square inch above atmospheric pressure), heat (in calories per square centimeter) and radiation are determined at various distances from ground zero. These distances are the radii of a series of concentric circles extending outward from the point of explosion. Within each circle, given the physical forces, the nature of the buildings and terrain, and the population concentration, it is possible to calculate the numbers killed and seriously injured.

The U.S. Arms Control and Disarmament Agency has made such calculations for every city in the United States with a population of 25,000 or more, at weapon sizes varying from 50 kilotons to 20 megatons.[2] In San Francisco, for example, the Agency calculates that a single one-megaton air burst would kill 624,000 persons and seriously injure and incapacitate 306,000. A single 20-megaton air burst would kill 1,538,000 and seriously injure 738,000.

These figures are serious understatements, however. They are based on a census population distribution, that is, they make the implicit assumption that everyone is at home, when in fact a population-targeted attack is much likelier to occur on a weekday during working hours, when the population is concentrated in central-city areas closest to ground zero. And they do not allow for the probability of a firestorm or mass conflagration as the secondary consequence of a nuclear attack. A firestorm—like those at Hiroshima, and at Dresden and Hamburg after conventional bombings during World War II—may burn for days, with ambient temperatures exceeding 800° centigrade. It increases the lethal area *five-fold.*[3] It also makes all conventional sheltering attempts worse than useless. At these temperatures, and with the exhaustion of oxygen supplies and the accumulation of toxic gases, shelters become crematoria. In Dresden and Hamburg, the only survivors were those who fled their shelters.

A 25 percent increase in the numbers of killed and seriously injured would be a conservative adjustment for these two factors. Thus corrected, the figures for San Francisco would be that:

- A one-megaton air burst would kill 780,000 persons (22 percent of the total Bay Area population) and seriously injure

382,000 (10.5 percent), for total casualties of 1,162,500, or 33 percent of the population.

- A 20-megaton air burst would kill 1,923,000 persons (53 percent of all the people in the Bay Area) and seriously injure 874,000 (24 percent) for total casualties of 2,797,000, or 77.4 percent of the population.

The figures illustrate the lack of precedent. There is no identifiable event in human history when a million people have been killed in one place at one moment. There is no previous situation in which there were 400,000 seriously injured human beings in one place.

The nature of these injuries further illustrates the magnitude of the problem. Among "survivors" there will probably be tens of thousands of cases of extensive third-degree burns. And in this kind of injury, survival and recovery depend almost entirely on the availability of specialized burn-care facilities, highly and specially trained medical and allied personnel, complex laboratory equipment, almost unlimited supplies of blood and plasma, and the availability of a wide range of drugs. No such facilities would remain intact in San Francisco; the number of Bay Area burn casualties would exceed by a factor of 10 or 20 the capacity of all the burn-care centers in the United States.

In addition to third-degree burns, hundreds of thousands of "survivors" would suffer crushing injuries, simple and compound fractures, penetrating wounds of the skull, thorax and abdomen, and multiple lacerations with extensive hemorrhage, primarily in consequence of blast pressures and the collapse of buildings. (Many of these victims, of course, would also have serious burns.) A moderate number would have ruptured internal organs, particularly the lungs, from blast pressures. Significant numbers would be deaf in consequence of ruptured eardrums, in addition to their other injuries, and many would be blind, since—as far as 35 miles from ground zero—reflex glance at the fireball would produce serious retinal burns.

Superimposed on these problems would be tens of thousands of cases of acute radiation injury, superficial burns produced by beta and low-energy gamma rays, and damage due to radionucleides in specific organs. Many would die even if the most sophisticated and heroic therapy were available; others, with similar symptoms but less actual exposure, could be saved by skilled and complex treatment. In practical terms, however, there will be no way to distinguish the lethally-irradiated from the non-lethally-irradiated.

Finally, this burden of trauma will occur in addition to all pre-existing disease among "survivors," and this list of problems is not based on consideration of the special problems of high-risk populations—the very young and the very old, for example—which are particularly vulnerable.

These are the short-range problems to which a medical response must be addressed. But who will be left to respond?

Physicians' offices and hospitals tend to be concentrated in central-city areas closest to ground zero. If anything, physicians will be killed and seriously injured at rates greater than those of the general population, and hospitals similarly have greater probabilities of destruction or severe damage. Of the approximately 4,000 physicians in San Francisco County, perhaps half would survive a one-megaton air burst; of the 4,647 hospital beds in the county, only a handful would remain. At 20 megatons, there would be only a few thousand physicians left in all of San Francisco, Alameda, Marin, San Mateo and Contra Costa Counties to try to care for 874,000 seriously wounded.

One carefully detailed study of an American city suggests that there would be 1,700 seriously injured "survivors" for every physician—and that includes physicians of all ages, types of training, states of health and location at the time of the attack.[4] If, conservatively, we estimate only 1,000 seriously wounded patients per surviving physician, if we further assume that every physician sees each patient for only 10 minutes for diagnosis and treatment, and if each such physician worked 20 hours a day, it would be eight days before all the wounded were seen—once—by a doctor. Most of the wounded will die without medical care of any sort. Most will die without even the simple administration of drugs for the relief of pain.

A closer look at these calculations reveals that they are absurd—and the absurdities have implications that extend far beyond issues of medical care.

Thus, the calculations assume that every surviving, uninjured physician would be willing to expose himself or herself to high levels of radiation. They assume that every physician will be able to identify the areas in which medical help is needed, get there with no expenditure of time, and find every one of the 1,000 patients. It is further assumed that physicians will spend no time on uninjured or mildly injured patients, on those with pre-existing illness requiring care, on those with acute illness unrelated to the bombing, or on those who merely believe they are injured, all of whom will demand his time and attention.

And all of this is happening in an area where there is no electricity, no surviving transportation system. What is left of the buildings is lying in what is left of the streets; the bridges are down; subways and tunnels are crushed; there is no effective communication system; there are no ambulances and no hospitals.

Finally, in each ten-minute patient visit, the "medical care" will be dispensed without x-rays, laboratory equipment, other diagnostic aids, supplies, drugs, blood, plasma, beds and the like. There will be no help from "outside." There will be no rational organization even of this primitive level of care. In short, this is not medical care at all, as we commonly understand it.

It is important to examine medical care scenarios not merely as an element in the essentially hopeless task of response to a nuclear attack, but as a metaphor for *all* complex human— that is to say, social—activities in the post-attack period. *What becomes clear is that all such activities require an intact social fabric*—not merely

the infrastructure of electric power, transportation, communications, shelter, water or food but the social enterprises, the complex human interactions and organizations supported by that infrastructure. That social fabric is ruptured, probably irreparably, by even a single nuclear weapon. Medical care is impossible in any real sense, not only because of the damage to the physical and biological environments, but most of all because it is a complex activity that requires a high degree of social organization.

The same is true of most other important human activities in complex urban societies. It follows that the only true meaning of "survival" is social, not biological. Simply to tally those who are still alive, or alive and uninjured, is to make a biological body-count that has little social meaning. The biological "survivors" in all probability have merely postponed their deaths—by days, weeks, months or at most a few years—from secondary attack-related causes. Life in the interim will bear no resemblance to life before a nuclear attack.

In the period from days to months after an attack, other problems of both social and medical significance will rapidly emerge.

Without functioning transportation, even assuming that effective social organization continues on the "outside," no food will come into the stricken area; remaining undestroyed stocks will be depleted rapidly. Extreme water shortages will occur almost at once. The average citizen of a modern American city uses between 50 to 150 gallons of water a day; in the post-attack period, a quart a day per survivor would be generous, and there will be no easy way to assure either potability or freedom from radioactive contamination.

Over the first two to four weeks after the attack, thousands of short-term survivors will die of radiation sickness, particularly of infection secondary to radiation-induced lowering of resistance. The problem of mass infection is particularly ugly. Even assuming that a firestorm conveniently incinerates 500,000 of the dead in a one-megaton attack, there will remain some 300,000 or more decomposing human corpses in the Bay Area. There will be no safe water supply or effective sanitation. The vectors of disease—flies, mosquitoes and other insects—will enjoy preferential survival and growth in the post-attack period because their radiation resistance is many times that of mammals. Most surviving humans will have reduced resistance to infection. It is hard to construct a scenario more likely to produce epidemic disease.

Finally, any *likely* population-targeted attack will assign many multi-megaton weapons to each major city, and therefore calculations based on a single one-megaton or 20-megaton strike are unrealistically conservative.

Other scenarios—the so-called "counterforce" exchanges aimed primarily at missile sites or various city-trading hypotheses—presumably would result in less *immediate* death and injury. But they pose medical and social problems of equal magnitude in the longer run, even if they do not almost automatically escalate into full-scale exchanges.

Mass evacuation of cities in a nuclear crisis, the current favorite of civil defense enthusiasts, would in itself be seen as provocative by an adversary and therefore would increase risks. According to testimony before a Senate subcommittee by representatives of the Federal Emergency Management Agency, effective evacuation would require "only eight days" from warning time to completion.

It is, once again, a technique aimed at short-term biological survival, not social survival: to what would the dispersed urban residents return?

The danger of nuclear war is a public health problem of unprecedented magnitude. It is not, however, unprecedented in *type*. There are many other medical problems to which a coherent response is not possible and for which there are no cures. One medical (and social) strategy is still available in such cases: Prevention.

1. Joint Committee on Atomic Energy, U.S. Congress, *Biological and Environmental Effects of Nuclear War: Summary Analysis of Hearings, June 22-26, 1959* (Washington, D.C.: U.S. Government Printing Office, 1959); *New England Journal of Medicine,* 266 (1962), pp. 1137-1144; S. Aranow, F.R. Ervin, V.W. Sidel, eds., *The Fallen Sky: Medical Consequences of Thermonuclear War* (New York: Hill and Wang, 1963); K.N. Lewis, "The Prompt and Delayed Effects of Nuclear War," *Scientific American,* 241:1 (1979), pp. 35-47; U.S. Congress, Office of Technology Assessment, *The Effects of Nuclear War,* OTA-NS089 (Washington, D.C.: U.S. Government Printing Office, 1979).

2. U.S. Arms Control and Disarmament Agency, "Urban Population Vulnerability in the United States" (Washington, D.C., 1979).

3. Lewis, "Prompt and Delayed Effects."

4. Aranow, Ervin and Sidel, *The Fallen Sky.*

14 Burn casualties

JOHN D. CONSTABLE

The crash of a partially filled 30-passenger airplane on Nantucket Island required the mobilization of all the emergency medical facilities of Greater Boston, a major surgical center. Yet we are seriously asked to contemplate and to discuss the possibility of ten thousand or a hundred thousand or even a million severely traumatized victims of a military nuclear explosion.

These numbers could be so matter-of-factly proclaimed only by those who are in complete ignorance as to the possibility of any adequate or even partially adequate medical treatment being made available to such survivors. As medical facilities are now set up, we can talk about how such injuries *should* be treated, but to transfer this knowledge to the practical possibilities of the treatment of the numbers of victims that have been predicted is categorically out of the question.

The injuries caused by a massive nuclear detonation would come from the various effects of such an explosion. Although I will briefly consider a number of different injuries, I would like to point out at once that burns or thermal injuries would be by far the most crushing burden on the available medical resources. In the case of very large explosions, radiation levels sufficient to cause immediate —or only very shortly delayed—death or massive radiation damage can be expected to extend not very far beyond the zone of lethal damage due to heat or blast. With relatively smaller explosions, such as those used in Japan, this is not strictly true. But it still means that among the survivors of an explosion of the size currently contemplated there would not be very many patients who might die soon from the immediate radiation effects.

The direct blast of a nuclear bomb will, of course, result in a number of injured survivors, but we must keep in mind that the explosion is relatively much more destructive to buildings than it is to persons. Whereas most ordinary houses are destroyed by an increase of perhaps five pounds per square inch in atmospheric pressure, the human body, as long as it is protected from injury by other objects, can withstand a temporary increase of 30 to 50 pounds per square inch.

There will, however, be very extensive traumatic injury to people within and around buildings, as a result of being blown out of them and by being damaged by debris from the destruction around them. Also, the initial blast effect of the explosion is characteristically followed by very powerful local winds rising to as much as 100 or 180 miles per hour and these, of course, will cause a number of severe traumatic injuries.

Most of those injured, whether they have been crushed, cut or blasted, but who have survived initial injury and have reached adequate medical facilities would, in most cases, be expected to require only one or two major surgical procedures. Although this might be very expensive in terms of time and material, including a great deal of blood and other support, the victims could then in most cases be expected to enjoy a relatively uncomplicated convalescence. The non-nuclear war experience in Vietnam has taught us that, at least among younger patients, even the most severe intra-abdominal or thoracic injuries, whether resulting from bombs, shellfire, or other causes, can be restored fairly quickly provided that the patient can survive until a medical facility has been reached and that major restorative procedures could then be carried out.

Of all the traumata resulting from a nuclear explosion, thermal injuries, even though heat and light contain only some 35 percent of the total energy of such an explosion, are first and foremost in terms of the extent of medical treatment needed in the first few weeks of injury. But, somewhat paradoxically, even though burn victims end up by consuming vastly more of the medical facilities than other injuries, most of those surviving burns can, in fact, be transported with minimal treatment for the first 8 to 24 hours after the injury.

Some special aspects of the burn problem need to be considered. How important is carbon monoxide poisoning? In an outdoor fire significant poisoning from this source is usually rare. Even when it occurs in a patient exposed to carbon monoxide in a confined space, either the levels of blood saturation have become so high that there is irreversible anoxic brain destruction, in which case there will be no recovery; or by the time the patient reaches a medical facility spontaneous recovery will be sufficiently advanced that the residual carbon monoxide absorbed will not be a major problem in treatment.

Patient anoxia resulting from most of the atmospheric oxygen having been used up by a fire—theoretically important in a so-called firestorm—is clinically rare. If the degree of thermal activity is sufficient to cause anoxic damage, then there will usually be concomitant fatal incineration. But if there is only a relative degree of anoxia, spontaneous recovery will occur by the time the patient has reached a medical facility.

Both carbon monoxide poisoning and fire-induced anoxia must be distinguished from so-called pulmonary burns, which remain one of the major largely unsolved therapeutic problems of thermal damage. This is a form of lung injury which usually takes from 24 to 72 hours to develop and is not, in fact, the result of direct thermal damage to the lung. If the heat around the patient's face is sufficient to result in actual destruction of the trachea, bronchae, or lungs, there will almost invariably be such devastating destruction of the face and other parts of the skin as well that the patient will not survive.

It is now generally accepted that the damage to the lungs is a result of the chemical activity of noxious products of incomplete combustion. Consequently, this type of burn is characteristic of fires in closed spaces rather than the open spaces which would be more common with a major bomb. Among people confined to buildings, pulmonary burns will be a major lethal factor. In the Coconut Grove fire in Boston some 40 years ago, over 400 people died, almost all without visible signs of burns. These deaths, which occurred mostly two, three and four days after the fire, resulted from pulmonary damage now believed to have been from the fumes from the plastic in the artificial palm trees and furniture coverings.

Although pulmonary damage may be a major cause of death in burns, it must also be recognized that this aspect of thermal damage does not really pose an immense burden on triage or on the medical system. This is because, even with the best possible facilities, it remains essentially untreatable. In general, these patients either have so major a lung injury that they will die, or with a relatively lesser degree of injury they will spontaneously recover quite quickly.

Aside from these secondary aspects, there will be two kinds of direct thermal injury from a nuclear explosion: one will result directly from the detonation; the other from the secondary fires following the ignition of available combustible material. These secondary fires are of at least two sorts. One is the possibility of a firestorm, and with the lower concentration of combustible materials in American towns and cities this is a little less likely than in many other parts of the world. Much more certain is the development of a major conflagration which would be essentially the sort of fire with which we are all too familiar, but enormously increased in scale. This fire would, of course, be associated with multiple smaller ones, starting from the breaking of gas mains, the failure of electrical pumps, the lack of water to put them out, and so on. The fires would presumably be spasmodic over a very large area.

Patients would thus be exposed to the risks of thermal damage from the bomb itself and from its secondary fires. I believe that there is no essential difference in the nature of burns resulting from these two etiologies. Burn damage to the skin results from a combination of the amount of heat and the time of exposure, these factors being very much modified by the presence or absence of clothing, the moisture content of the atmosphere, and other factors. An explosion results in an almost instantaneous exposure to a very high heat level with damage occurring over an incredible distance; but the nature of the injury is not, I think, different from other forms of thermal burns. It simply means that there can be much more severe damage in a very short time if the heat to which one is exposed is very great.

I must at least mention the problem of thermal injuries combined with the effects of radiation. All patients seriously injured by nuclear explosion who have also had a significant amount of radiation injury will be more difficult to treat. My assumption here is that relatively few surviving patients will have received sufficient radiation to result in death within a matter of weeks or months from the radiation alone. But even with those who have received smaller doses of radiation, the damage to the immune system and to blood element regeneration results in the patient's being more prone to invasive sepsis, in less satisfactory healing, and in an increased risk of death from a thermal injury which might otherwise not have been fatal.

Experimental studies have shown that a burn from which a normal animal can be expected to recover becomes lethal if the animals have been previously or concomitantly exposed to nonlethal radiation. (A medically interesting note: in dealing with a very small number of victims, as in a nuclear reactor explosion, the immunosuppressive aspects of radiation might not be totally disadvantageous. In current practice, severely burned patients are treated by immunosuppression in order to allow for the extensive use of allografts.)

First degree burns are at their very worst equivalent to a severe sunburn. They may result in some transient dehydration, certainly considerable pain, but under any emergency conditions, these require essentially no treatment and must be considered of no particular medical consequence.

Second degree or partial thickness burns (the latter term is much to be preferred) are, from the point of view of the immediate surgical problems, almost as severe an injury as are full thickness burns. A deep partial thickness burn requires essentially the same amount of resuscitative effort, the same difficult nursing, the same elaborate dressings, and the same extensive care during the first three to four weeks. Although these injuries heal from the base and therefore no skin grafting is required, and the eventual problems of resurfacing the patient are a great deal simpler, the immediate problem of care is almost as great as with a full thickness burn. The two groups should be combined from the point of view of trying to evaluate the early load on the medical system.

It is very difficult to estimate accurately the extent and number of burn survivors in a population exposed to a nuclear explosion. The figure might vary by as much as a thousandfold, depending upon specific factors prevailing at the time of the explosion. These include not only the size of the bomb and the above-ground level of the explosion, but also the atmosphere. Even a moderate degree of opacity in the air strikingly reduces the range of thermal damage. Other factors include the season, the time of the day, and the extent to which the population has been pre-warned. These conditions partly determine the amount of clothing being worn and whether people are outdoors or not, since at the periphery of an explosion protection from, or at least reduction of the extent of, thermal injuries can be fairly easily obtained.

Even with these caveats and modifiers, it has been estimated that for a one-megaton nuclear explosion, with ten-mile visibility, only first degree burns might be expected within a seven-mile radius; second degree or partial thickness burns within a six-mile radius; full thickness destruction within five miles. If the atmosphere were sufficiently opaque to reduce visibility to two miles, then the second degree zone would be reduced from six miles to something under three and the others changed proportionately.

The two-and-a-half mile radius from the center of the explosion —the limit of second degree burns if the atmosphere restricts visibility to two miles—is approximately the same limit as that of five pounds per square inch of blast. This is generally considered to be the lethal average for humans, due to secondary effects of blast and wind, so it becomes clear that unless the atmosphere is even more opaque, the greatest number of severely damaged survivors will be within two-and-a-half to six miles from the center and their trauma will be the result of thermal injury rather than other causes.

Unfortunately, this is the form of trauma which demands the largest amount of medical assistance if it is properly managed. There is, indeed, no injury that can be counted on to use up more hospital facilities than can a severe burn. Triage—judging which burn patients will survive—would be very difficult and it may optimally be necessary to treat a great many patients for extended periods who will eventually die from their injuries.

The burn literature has been filled over the last ten years with reports of progress in salvaging the severely burned. Many new methods of infection control have come into use, including various surface antiseptic agents and topical antibiotics. The surface control of infection has prevented the conversion of partial thickness to full thickness burns by sepsis and has strikingly improved overall results in burn salvage. There has also been much effort to control systemic infection, both by the use of antibiotics and by elaborate isolation techniques. There are life islands in which patients are more or less isolated in a plastic enclosure, and more recently laminar flow units. These latter are devices in which the air is regularly replenished and replaced so that bacteria are swept away and the air is kept essentially sterile. All of these methods have helped reduce death from infection.

Another recent development is the early surgical excision of burns. This is now often done, and although it is usually not safe to excise more than one-fifth of the patient's body surface at one sitting, surgery may be carried out on the first or second day after burn, and with maximum support, again on the fourth, and so on, ending with as much as 80 to 90 percent of the skin being excised. Massive excision has been combined with immunosuppression so as to allow for the use of typed allografts taken from living donors or cadavers. It is possible with these methods to obtain some dramatic results although they are still cosmetically or aesthetically relatively grotesque. These are certainly very satisfying to the burn surgeons involved, and reasonably so, perhaps, to the patients and their families.

It is absolutely essential to recognize, however, that any really severe burn that is salvaged may require as many as 30 to 50 operative procedures, both immediate and delayed, and months and months of hospitalization. This imposes immense strains on the medical facilities available. With the newer and more dramatic methods, there is at least the possibility, if sufficient material and personnel are poured in, of salvaging burns in the 85 to 90 percent range.

This, of course, makes triage much more difficult. We would be faced with an enormous group of patients sustaining 20 to 90 percent burns who might survive if treated. Except for localized burns of the hands and face, I exclude burns affecting under 20 percent of surface, because most of these can be relatively easily treated. What is, in fact, involved in the possibility of treating large numbers of severe burns?

Some years ago the Shriners of North America, who had for years donated large sums to look after orthopedically crippled children, began to have less orthopedic demand because of diminishing polio, tuberculosis, and chronic osteomyelitis. They therefore became interested in building specialized burn hospitals for children. Their plan was to start with three burn units and then expand, possibly adding another 15 or so to match the number of orthopedic hospitals they were already maintaining. These initial three units were built in Boston, Galveston and Cincinnati. In the 15 years since these three 30-bed hospitals were built, it has not been practicable to build even one other unit, because the three burn units, with a total of 90 beds, use up a budget similar to that of 19 orthopedic hospitals, most of which are of comparable size.

The cost of running a single 30-bed hospital, in which half of the beds are reconstructive and where there would rarely be more than ten acute burn cases at one time is in the neighborhood of $4 million per year. There are, all over the United States, something like 1,000 so-called burn beds. These are in specialized institutions willing to look after severe burns, but to do this appropriately each burn patient requires specialized individual nursing for quite a long time. At most, one nurse can look after two patients.

Severe burn cases require not just one major operation, but may need general anesthesia every other day and regular trips to the operating room for weeks or even months. There are elaborate dressings and the application of appropriate antibiotics or at least antiseptic agents. The patients require large amounts of blood, albumin, and other human blood derivatives. They may need enormous areas of allografts, and even in wartime it may be difficult to obtain sufficient quantities of these from cadavers.

Whereas most traumatic lesions are more or less definitively treated immediately, and the victims either recover or die, burns are peculiar. The burn patient is not so ill during the first 12 to 24 hours. I have seen a number of older patients with 40 to 50 percent full thickness, clearly fatal, burns who, for the first 12 to 24 hours after their injury, appeared in reasonably good condition. They were capable of consulting their lawyers and doing whatever needed to be done. It is after this initial period that the patient becomes sicker and sicker, and this critical hovering between survival and death may go on for weeks or months. Then, once a burn has been initially resurfaced, it may need months or years of reconstruction. And even with all of this, anyone who is discriminating or humane would recognize that the end results are indeed pathetically poor.

It is very difficult to estimate the cost of such cases in dollars because, to the best of my knowledge, no health program or insurance pays adequately for burn care. Blue Cross/Blue Shield and similar programs admit that they cannot afford to pay the true cost. Nonetheless, it is reasonable to put the cost at anywhere from $200 to $400 thousand for the average severe surviving burn case.

Even though there are 30-bed burn units, such as the Shriners' or those at large general hospitals, they can, in fact, handle only two or three fresh severe burns at once. If there is a large group of such burns in a major accident, they will have to be distributed for any effective treatment.

Even the most famous burn disasters of recent years—the Coconut Grove, plane crashes, or the Hartford Circus—have resulted in very few initially surviving major burns, but the expectation of any holocaust, such as a nuclear bombing, is that there will be at least thousands of severely burned people immediately surviving. The most conservative calculation of the thermal injuries resulting from an isolated one-megaton or "minimal nuclear explosion," with preservation of all U.S. medical facilities and the availability of immediate and perfect triage and transportation, will completely overwhelm what we consider to be one of the most lavish and well-developed medical facilities in the world. It is impossible to imagine the chaos that would result from a larger attack in which the hospitals themselves were partially destroyed and there was no ability for significant triage or intercenter transportation. The medical facilities of the nation would choke totally on even a fraction of the resulting burn casualties alone.

15 Infection and communicable diseases

HERBERT L. ABRAMS

No arts; no letters; no society . . . and the life of man, solitary, poor, nasty, brutish and short.—Hobbes, 1651.

Hobbes' description might well embody the primitive life that would follow the devastation and chaos of a massive nuclear war. The potential for regression to a social structure unknown to Western man has been appropriately emphasized. But the nature of the medical problems confronting survivors has not been widely conveyed or fully understood.

Perhaps the most logical approach is to view this matter in sequence, from the moment of attack through the periods that follow. Within that framework, the characteristics of each period can be delineated, although the duration cannot be precisely stated. The type and severity of the initial injury will clearly affect and interact with the response to new medical problems as they arise. Finally, each problem may be expected to extend beyond the period in which it is initially encountered (Table 1).

The effects of burn, blast and radiation have dominated discussions of the post-attack period. But in the intermediate term, infection and the spread of communicable disease represent the most important threat to survivors.

The United States we will view will have undergone a 6,559-megaton attack. The magnitude of such an attack—the so-called CRP-2B model used by the Federal Emergency Management Agency in civil defense planning—reflects the world of the 1980s [1, pp. 20-21]. In terms of yield, it represents 524,720 Hiroshima

Table 1

Medical problems during the attack and post-attack period

Problem[a]	*Immediate effects*	*Shelter period*		*Post shelter survival*	*Long-term effects*	
	1st hour	*1st day*	*1st 4 weeks*		*Recovery period*	*Future generations*
Flash burns	+					
Trauma	+					
Flame burns and smoke inhalation	+	+				
Acute radiation	+					
Fallout radiation	+	+	+	+		
Suffocation and heat prostration		+	+			
General lack of medical care		+	+	+		
Dehydration			+			
Communicable diseases			+	+		
Exposure and hardship			+	+		
Malnutrition			+	+		
Cancer					+	
Genetic						+

[a] In approximate order of time in which inflicted.

bombs. The targets of attack, in order of priority, will have included:

- military installations;
- military-supporting industrial, transport and logistic facilities;
- other basic industries and facilities which contribute to the maintenance of the economy; and
- population concentrations of 50,000 or greater.

Some 4,000 megatons will have been detonated on urban areas and population centers.

Moments after the attack 86 million—nearly 40 percent of the population—will be dead. An additional 34 million—27 percent of the survivors—will be severely injured. Fifty million additional fatalities are anticipated during the shelter period, for a total of 133 million deaths. Many of the millions of surviving injured will have experienced moderate to high radiation doses. Approximately 60 million may survive and emerge from the shelter period without serious injury and with relatively limited radiation exposure. The periods under consideration are:

Immediate effects. During the barrage period, the explosions almost instantaneously inflict millions of lethal and non-lethal blast, thermal and immediate radiation injuries on those caught in and around the blast areas.

Shelter period. From the time of the attack to days or weeks later, those surviving the initial explosions attempt to sustain themselves in fallout shelters, amid intense radiation, fires and deprivation.

Post-shelter survival. Fallout has reached an "acceptable" level for emergence for variable times. The problems of obtaining food, developing independent shelter and recovering from the acute injury must be confronted. The injured must be nursed, the dead buried, debris cleared, the harvest reaped and the next harvest sown. In a primitive and hazardous environment, survival is the only coherent goal.

Long-term effects. Survival has been accomplished and some kind of recovery initiated. A societal structure is emerging, food supplies have been secured, shelter obtained, and communities established. A primitive social organism, with potentially intense regional and intra-regional competition for food supplies, faces all the problems of underdeveloped countries. During the early years, the first cases of radiation induced leukemia appear; later the solid cancers will develop in the lungs, thyroid, breast and colon.

The problem of infection and communicable disease must be addressed during the shelter period and more particularly during the post-shelter survival period. It cannot fully be separated from shelter conditions (crowding, poor ventilation), food, water, antibiotics, burn, blast residua or radiation and its effects.

The skeptic may inquire whether the deaths of a few million more people really matter in the post-attack world. One important study omits possible epidemics in its fatality estimates on the grounds that "with proper planning, the overall effect. . . would be small compared with the short term effects of the attack" [2]. Others suggest that epidemics may demand more than a footnote and are, in fact, "a matter of considerable uncertainty" [3, p. 140]. An overriding reason for considering the problem carefully inheres in its potential interactive impact on the recovery process.

What is the nature of the threat, and why is the likelihood of infection so much greater in the post-attack world?

Increased susceptibility to infection

Most survivors will be more susceptible to infection both because of the pervasive direct effects of nuclear weapons and the subsequent pressures and hardships confronted. Several factors will be of special importance.

Radiation. Radiation affects the immune system in a number of different ways, not least of which is its capacity to injure the bone marrow and the lymph nodes. Significant hematologic changes may occur with doses as low as 50 rems. The following changes occur:

- decreased antibody response (Altered immunity following total body irradiation lasts from weeks to months. Some changes may be of far longer duration following total nodal irradiation.),
- increased susceptibility to toxins,
- decreased effectiveness of cellular defense mechanisms and
- decreased effectiveness of immunizing agents [4].

Thus, vaccination will be more dangerous and less effective, and antibiotics may well induce reactions.

Radiation also has a major effect on the lining or mucous membrane of the intestine. "Ulceration . . . spreads through the entire gastrointestinal tract. . . . The multiplication of bacteria, made possible by the decrease in white blood cells and injury to other immune mechanisms of the body, allows an overwhelming infection to develop." [5]. The ulcerated mucosa provides a portal of entry into the blood stream for gram negative organisms, with bacteremia as a certain sequel.

Federal estimates indicate that 23 million survivors of a large-scale nuclear war will suffer from radiation sickness, implying a mean dose of 200 rems or more [6, p. 62]. But the number of cases with doses between 100 and 200 rems will probably equal that, so that 50 percent of the population might have lowered resistance to disease from radiation exposure alone [7].

Trauma and burn casualties. Among the millions suffering from trauma and/or burns, over a third will also suffer radiation sickness [6]. Aside from the risk of infection related to open wounds, weakness and incapacitation may be expected to increase their general vulnerability. There is also a known synergy between burns and radiation that profoundly increases the mortality rate. The organisms that will infect the burn sites include pseudomonas and serratia, either of which may be troublesome in patients with normal immune mechanisms, and both of which will be difficult to eradicate in an immuno-suppressed population.

Malnutrition and starvation. During the shelter period, available quantities of food would vary. Adults can maintain health for several weeks with only minimal amounts of nutrition [8]. In some locales, fallout might prevent emergence from shelter to a point where health would deteriorate. Infants, young children, and pregnant women, in particular, might experience severe malnutrition due to insufficient or inappropriate foods during an extended shelter stay [9, p. 79]. Furthermore, although survival with little food is possible in the resting state, any heavy physical exertion can only be tolerated with caloric replacement.

During the post-shelter phase, three crucial periods can be defined: shelter emergence until the first harvest, first harvest until the next harvest, and all subsequent seasons. The first two periods depend a great deal on the time of year of the attack, and the third depends on post-attack recovery and environmental conditions. Upon shelter emergence, most of the food stores will be destroyed in urban areas; other supplies will most likely be consumed during or soon after the shelter period [1, p. 109].

The essential food supply that may be available will be the grain stored on and off farms in small towns and rural areas. The lives of millions of survivors would depend upon this supply. Since the amount of grain stored on and off farms varies considerably over a given year, the supply could last from 200 to 500 days, with great dependence on the next harvest [1, pp. 110-15].

But food piled high in silos in remote regions will do little good for a hungry populace. Grain must be obtained, transported, and distributed to the survivors where they are located. Grain transportation will be the most important survival activity in the immediate post-shelter period. This will be made much more difficult by the negative correlation in the United States between population and grain density [10].

Before grain can be transported to populated regions it must be procured. In the uncertain post-attack environment, food will take on an immense value, and those persons or regions with food supplies might be reluctant to give them up.

"It is not at all clear that many localities would be willing to part with large stockpiles of food in the post-attack period . . . attempts to requisition food post-attack could lead to local armed conflicts" [11, p. 30].

Furthermore, assuming that trucks and highways will be usable, an essential commodity for the transportation of grain is fuel. It is estimated that as much as 99 percent of U.S. refining capacity could be destroyed [1, p. 153]. Liquid fuels would be in "the most critical situation," and there could be bitter tradeoffs. Agriculture will be a large user of fuel. The only possible choices might be between hunger today through lack of grain shipments, and hunger tomorrow through poor agricultural production. In many populated parts of the country, famine would be a reality.

The first post-attack harvest would depend on many factors, including the time of year of the attack. A winter attack would do the least damage. Fallout would have largely decayed to safe levels by the time of the next planting. In early spring, those plants above the ground would probably be killed by fallout, but might be replanted in time. An attack in June might destroy up to 70 percent of the U.S. crop. In fall, the plants would be grown, and thus less vulnerable to radiation. But an attack prior to the harvest might prevent the farmers from harvesting on time [3, p. 108].

A reasonable level of nutrition is essential to maintenance of control of infection. Malnutrition lowers physiological resistance to disease and heightens susceptibility to pathogenic organisms [12].

Dehydration. It is estimated that a healthy person in the hot environment of a fallout shelter would need about a gallon of water per day to prevent dehydration [9, p. 99]. Furthermore, the incidence of dehydrating diarrhea and vomiting is expected to be high among survivors of a nuclear attack. But no water is normally stored in public fallout shelters, and no means by which to transport or store water will necessarily be available. In a fallout shelter containing 1,000 persons, it would be difficult to obtain thousands of gallons of water daily for a sustained period in the rubble-strewn wasteland of a blast area. For that third of the population in areas where initial radiation was over 3,000 rems per hour it will be at least five days before radiation levels allow shelter breaks of up to one hour per day [1, pp. 39, 61-63]. Some may die of dehydration in the shelter; survivors will be profoundly weakened and far less able to resist infection. When water is available, there will be great difficulty keeping it uncontaminated.

Exposure and hardship. Widespread destruction of urban housing will occur, with major damage to rural housing as well. Heating fuel may be unavailable. General hardship will be accompanied by the need for intensive labor, with exhaustion, fatigue and poor nutrition promoting great vulnerability to infection.

Lowered natural resistance to disease. Surviving Americans would for the first time experience the underdeveloped world as their habitat. Unlike the population of impoverished lands, however, Americans do not have the high natural immunity to a host of dangerous diseases that allows many in the Third World to survive. The omnipresence of antibiotics has clearly altered the normal production of antibodies to infectious agents among the developed nations [9, p. 99]. With potential destruction of the pharmaceutical industry, as well as post-attack disorganization and chaos, antibiotics will be in short supply for countries that have depended on them.

The successful campaigns to eliminate the epidemics of lethal diseases, such as cholera and typhoid fever, have been accompanied by a failure to develop resistance to these diseases. Even with our past vaccination policies, 14 percent of the population was thought to be unprotected from smallpox [13]. This may be academic if smallpox is truly eradicated, as some have claimed. But a store of smallpox virus remains at the Center for Communicable Disease in Atlanta (a precaution should vaccine be required). If the Center were to experience the blast of a nuclear weapon in the region, the virus might soon be out of control. Reintroduction of such "exotic" diseases might find the population incapable of handling them, as were American Indians when exposed to diseases of Europeans. Measles, whooping cough and diphtheria may be rampant in unimmunized infants, and beta hemocytic streptococcal infections will be widespread.

Increased spread of disease

In a post-attack situation, five factors would be critical to increasing the spread of disease:

Shelter conditions. Large public shelters may operate under severe space limitations, with thousands packed into inadequate areas. Under these conditions, hepatitis and respiratory and gastrointestinal infections could spread rapidly. Most shelters will lack adequate forced ventilation. When ventilation systems *are* present, the blowers and fans could easily be rendered inoperable by an overpressure of one pound per square inch and the systems blocked by electric power failure [9, p. 40]. Heat and humidity in the shelter would then increase and the absence of a continuous flow of fresh air would encourage the spread of infective microorganisms.

The length of time that fallout radiation will enforce basement and underground occupancy will be an important determinant of disease spread. This period may well be one week to several months [14]. Those in highly radioactive cities will have much longer stays. Even after it is permissible to work outside, it may still be necessary to eat, sleep, rest, and so forth in the fallout shelters. "The spread of respiratory and other diseases . . . would be difficult to control in long-occupied shelters" [9, p. 95].

Sanitation. The barriers to communicable disease spread today —sterilized water, properly prepared and refrigerated food, sewage treatment and waste disposal—will be seriously compromised in the post-attack environment.

With contaminated water and food, along with generally unsanitary conditions, a host of enteric diseases not yet experienced by most Americans would probably spread widely. These include infectious hepatitis, salmonella, E. coli, amebic dysentery and possibly typhoid and paratyphoid.

Insects. Insects are generally more resistant to radiation than is man. This fact, and the existence of corpses, waste, lack of sewage treatment, depletion of avian predators and destruction of insecticide stocks and production would engender a huge increase in insect growth.

> "Mosquitoes would multiply rapidly after an attack. . . . The fly population would explode. . . . Most domestic animals and wild creatures would be killed. Trillions of flies would breed in the dead bodies" [9, p. 94].

Insect-borne diseases are more amenable to control than are respiratory or enteric diseases. But the absence of control of insect growth, combined with failure to provide adequate sanitation, might sharply limit the capacity to control such diseases as typhus, tularemia, Rocky Mountain spotted fever, yellow fever, malaria, dengue fever, encephalitis and anthrax [13].

Corpses. The health threat created by millions of post-attack corpses is a serious one. In many areas radiation levels will be so high that corpses will remain untouched for weeks on end. With transportation destroyed, survivors weakened, and a multiplicity of post-shelter reconstruction tasks to be performed, corpse disposal will be remarkably complicated. In order to bury the dead, an area 5.7 times as large as the city of Seattle would be required for the cemetery.

Animals. Like man, domesticated animals such as cats and dogs, present in huge numbers in populous areas, will experience altered immunity. Diseases of dogs such as tuberculosis, brucellosis and leptospirosis may spread from animals to surviving men. Cats and dogs running wild or injured will feed on carrion and be exposed to swarms of flies or other insects. Rabies infects not only dogs and cats but also raccoons, foxes and skunks. In disintegrating areas, wild rabid animals may become a major hazard.

Decreased capacity to respond to infection

Effective response to infection would be limited by various factors:

Government organization. The United States has developed an extraordinary ability to take effective countermeasures against communicable diseases. Should an outbreak occur, a public health network will be informed, the rest of the country alerted, and appropriate steps taken. In 1947 a man infected with smallpox mingled with New York City crowds for several days. More than 6,350,000 persons were immediately vaccinated; as a result, only 12 additional cases appeared [15]. More recently, the unfortunate swine flu episode illustrated how the hint of an epidemic could bring enormous medical resources to bear upon the threat.

But coherent efforts to control and limit the spread of disease require a modicum of surviving government, organized geographic units, communication networks and a favorable enough survival situation so that physicians and health officials can perform their tasks.

All of these conditions are speculative in the post-attack world. Most radio contact will be eliminated by nuclear weapons effects [16]. Treating the wounded will require the full attention of available medical resources. For a sustained period, surviving officials may have to remain in shelters, with food and water their primary concern. The huge number of injured, the tenuous food situation, massive industrial destruction, enormous debris removal and body disposal tasks, and disparity between food rich and food poor regions will seriously undermine interregional cooperation. [11, p. 30].

Disease detection and diagnosis. Health countermeasures against potential epidemics depend upon the involvement of physicians. Casualties among physicians and other health personnel will approximate 80 percent. This is higher than the casualty percentage of the population as a whole (73 percent) because physicians are disproportionately represented in the large cities. Government estimates suggest that there may be 79,000 uninjured physicians surviving a large-scale attack. They will have 32 million injured to treat, of whom 18 million suffer from radiation sickness and 14 million from trauma and/or burns. If only the trauma and burn victims are included, this represents a ratio of 175 injured for every uninjured physician [6, p. 62]. But many of the injured patients will be highly concentrated in specific locales, so that the ratio will be far greater in the most damaged areas. Moreover, physicians will be confined to shelters for many days or weeks. If we assume that the average shelter contains 100 people, and no more than one physician is in any shelter, only one in 12 shelters will have an uninjured physician [11, p. 72]. If there should be a serious epidemic, it will affect physicians as well. More physicians would become incapacitated as the number of sick increased, further raising the injured-to-physician ratio.

Laboratories are essential for dealing with communicable disease, but they will be highly vulnerable to the effects of an attack. Of the 50 State Public Health Laboratories and six Federal Communicable Disease Centers, only 11 are in low-risk areas. None of the surviving laboratories will be able to provide tests for amebiasis, viral gastroenteritis, hepatitis-A, or plague [17, p. V-32].

The prompt detection and diagnosis of communicable diseases essential to identification of medical resource needs will thus be severely impeded during a crucial period.

Inadequacy of countermeasures. Many of the countermeasures required for enteric and vector-borne diseases will be unavailable in the disorganization of the post-attack period. Their control involves assuring supplies of pure water and uncontaminated food, disposal of sewage and waste, and removal of breeding areas for insects and rodents (Table 2). The spread of respiratory disease will be greatly enhanced by the crowded quarters and poor ventilation of fallout shelters. Both antibiotics and immunization would play an essential role in stemming epidemics.

But how effective would they be?

Antibiotics. Antibiotics are ineffective in combating viral disease and cannot limit the spread of such infections as smallpox, viral gastroenteritis, and influenza. Several dangerous bacterial diseases such as diphtheria and tetanus respond poorly to antibiotics. [13]. Furthermore, the demand for antibiotics will be

Table 2

Principal communicable disease countermeasures by mode of transmission

	Mode of transmission		
Countermeasure	*Enteric*	*Person-to-Person*	*Vectorborne*
Antimicrobial therapy	+	+	+
Excreta disposal	+	-	-
Food hygiene	+	-	-
Immunization	+[a]	+	-
Potable water	+	-	-
Public information	+	+	+
Vector control	+	-	+

Source: D.R. Johnston, M.N. Laney, R.L. Chessin and D.G. Warren, *Study of Crisis Administration of Hospital Patients* and *Study of Management of Medical Problems Resulting from Population Relocation,* RTI/1532/00-04F, prepared for Defense Civil Preparedness Agency, (Research Triangle Institute, Sept. 1978), p. v-32.

[a]Typhoid fever only.

large. If laboratory tests are not available, they will be prescribed for all undiagnosed ailments. The effort to use them prophylactically in those exposed to high doses of radiation will be widespread. Stores of antibiotics will be largely destroyed in urban centers, and those still intact may be inaccessible for days or weeks because of intense fallout radiation. Rural stocks may be more plentiful because of farmers' stockpiling for livestock and might be utilized to prevent human disease in rural areas. While it would be imperative for post-attack production of antibiotics to resume immediately, it is highly probable that the pharmaceutical industry would be "virtually eliminated" in a massive attack [18; 19]. The strictest rationing of antibiotics would be essential so that they could be available when most needed.

Immunization. For several hazardous diseases, such as tetanus, poliomyelitis, measles, influenza, and typhus, immunization is the only effective direct means of control. In post-attack conditions, however, the effectiveness of vaccination programs will be diminished by the impact of radiation on the immune system. Millions who have had substantial radiation doses and will therefore most need immunization will benefit least. Because vaccines are effective only for specific diseases, or specific strains, accurate diagnosis will be essential but may be impossible in the absence of adequate laboratory facilities. If an unfamiliar disease or strain emerges, existing supplies will be useless in combating its spread. Production of the specific vaccine in quantity would be difficult, if not impossible.

Studies in the late 1960s identified 23 diseases that might be significant in the post-attack environment [17, pp. V 6-8]. Many of these are encountered in endemic form throughout the country (Table 3). Among them, potential epidemic sources may be divided into two categories (Table 4). The first includes the classic epidemic diseases, fortunately of low incidence; the second, diseases of heightened incidence but low mortality [20, pp. 205-6]. Respiratory diseases, including viral pneumonias, influenza, pneumococcus and streptococcus infections and tuberculosis will particularly affect those living in crowded blast or fallout shelters, with an augmented impact on the young and the old. The diarrheal diseases such as salmonellosis, shigellosis, campylobacter and viral gastroenteritis will be widely prevalent. Although their mortality rate is usually low, in the presence of radiation injury to the gastrointestinal tract it will be substantially increased. Furthermore, these diseases, as well as infectious hepatitis, may spread rapidly in the absence of adequate sewage disposal, pasteurized milk, or appropriate sanitary precautions. The group of diseases endemic

Table 3
Infectious diseases in the post-attack period

Communicable diseases of potential post-attack significance	*Reported cases in the United States in 1979*
Amebiasis	4,107
Diphtheria	59
Encephalitis, arthropod-borne, viral	1,266
Botulism food poisoning	45
Salmonellosis food poisoning	33,138
Hepatitis, A	30,407
Influenza	0.3[a]
Measles	13,598
Meningococcal meningitis	2,724
Plague	13
Pneumonia	19.7
Rabies	4
Shigellosis	20,135
Smallpox	0
Tuberculosis	27,669
Typhoid fever	528
Typhus fever	69
Whooping cough	1,623

Source: Center for Disease Control, "Annual Summary," *Morbidity and Mortality Weekly Report,* 28 (Sept. 1980), p. 3.

[a]Death rate per 100,000 in 1979.

to rural sections, and thus a danger to evacuated populations, includes rabies, plague and tetanus. A number of other diseases such as cholera or influenza might spread rapidly in devastated areas [21].

A more detailed view of two among many diseases—tuberculosis and plague—that are generally considered well controlled in Western society will indicate the roots of the concern for the role of communicable disease in the transformed post-nuclear world.

Tuberculosis

The "Great White Plague" of the nineteenth century was a lethal infection for large segments of the population. Death rates ranged as high as 550 per 100,000 in New York City [22, p. 50]. If the annual U.S. death rate of 184.7 per 100,000 from tuberculosis in 1900 to 1904 characterized our present population of 225 million, all the U.S. deaths from World Wars I and II, Korea, and Vietnam would be outnumbered in one year and 10 days [23].

Should this concern us for the post-attack period, when we know that the mortality rate of tuberculosis has fallen below 1/200th of the 1900 to 1904 figures? In 1978 there were only 2,830

Table 4

Potential epidemic diseases

Epidemic diseases of low incidence	*Serious diseases of higher incidence*
Cholera	Diarrhea
Malaria	Hepatitis
Plague	Influenza
Shigella	Meningitis
Smallpox	Pneumonia
Typhoid fever	Tuberculosis
Typhus	
Yellow fever	

deaths, and 28,521 new active cases in the U.S. The percentage of the population who are positive reactors has also dropped dramatically to 4 to 8 percent of all tested [24].

But the bulk of this decline was achieved without the aid of drug therapy. By 1944, when the modern era of effective antituberculosis drug treatment began [25], the mortality rate had dropped to 43.4 per 100,000, less than 24 percent of its 1900 to 1904 rate [22, p. 811]. This change was largely attributable to improved socio-economic circumstances, particularly since the incidence and mortality of tuberculosis rose and fell frequently in the past with altered societal conditions, especially in times of war (Figure 1).

In World War I, mortality increased 218 percent in Warsaw, reaching a ratio of 974 per 100,000 in 1917. Belgrade had the highest incidence of all European cities; the rate in 1917 reached 1,483 per 100,000. This represented almost 1.5 percent of the population dying from tuberculosis per annum [26, pp. 10-14]. During World War II, the death rate rose 268 percent in Berlin, 222 percent in Warsaw, and 134 percent in Vienna. An analysis of 2,267 chest roentgenograms at the Dachau concentration camp at the time of liberation showed that 28 percent had evidence of tuberculosis, of which nearly 40 percent were "far advanced" [26, pp. 14-18].

All of the factors that would increase susceptibility to disease and spread infection in general would be particularly applicable to tuberculosis. The destruction of housing, lack of fuel, shortages of food, medicine and clothing, and sustained periods of labor and struggle would create precisely the setting in which tuberculosis has flourished in the past.

- *Crowding.* Crowded living conditions for survivors are likely to become the *norm* for the U.S. population after a massive nuclear war. This will begin in the shelter period, with most people

Fig. 1. Tuberculosis mortality and first-class protein intake for England and Wales, 1938 to 1946 (1938 figures taken as 1)

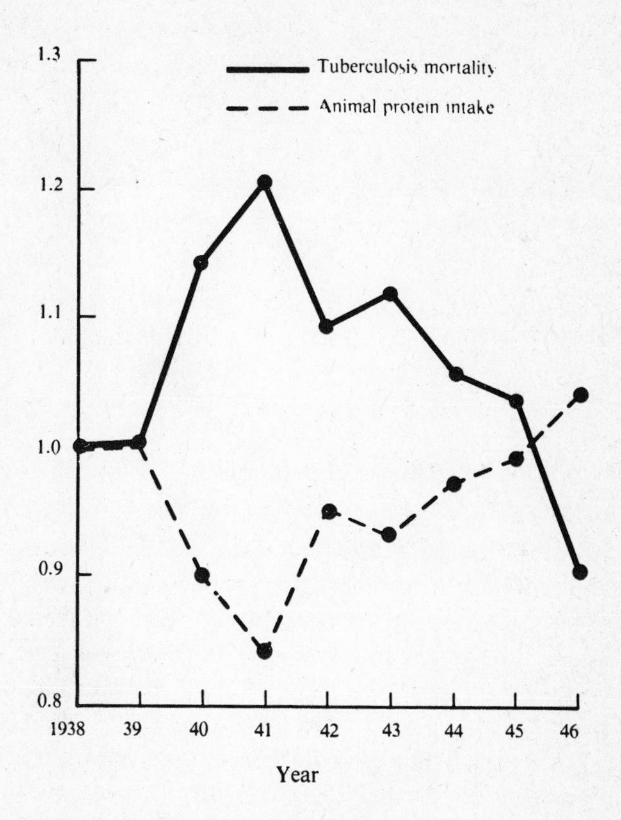

Source: Adapted from H.H. Mitchell, *The Problem of Tuberculosis in the Postattack Environment,* RM-5362-PR, The RAND Corporation (June 1967), sponsored by the United States Air Force.

remaining in shelter two weeks or more to avoid intolerable radiation. It will continue during the post-shelter phase because of the destruction of housing. Under these circumstances, tuberculosis can spread rapidly, with fulminant outbreaks in nursing homes, schools, prisons and aboard ships. In 1959, for example, an outbreak occurred aboard a U.S. Navy destroyer. Despite vigorous efforts at control, 26 percent of the ship's 236-man complement acquired the disease during an 18-month period [27].

• *Exposure and stress.* Stress, fatigue, exposure, and hardship will produce a work force that will include infirm, weakened and sick individuals. Relapse of latent cases and susceptibility to new infection will be widespread.

• *Undernourishment.* Nutritional status has been traditionally associated with increased incidence and mortality from tuberculosis. Over 20 different studies have shown the relationship between food quality and tuberculosis, most striking during wartime [28; 29].

Although animal protein is particularly important (Figure 1), the post-attack diet will consist mainly of grains and beans [30].

• *Tuberculosis among "virgin" populations.* Fewer Americans have been exposed to tuberculosis today than ever before. There are numerous examples of catastrophic epidemics of tuberculosis among largely unexposed populations. South African troops in World War I had a mortality rate of 1,745 per 100,000 from tuberculosis, while British troops had a rate of only 11 per 100,000. When Saskatchewan Indians were removed from nomadic to reservation life around 1880, their death rate from tuberculosis reached 9,000 per 100,000 [26, pp. 48-53]. In the post-attack period, the disease may be particularly virulent among Americans because of the lack of protective antibodies.

• *Radiation.* Studies in Hiroshima showed no increase in tuberculosis a few years after the bombing, but the data on the immediate and intermediate post-bombing period are unclear [31]. It is known that immuno-suppressive therapy is an important risk factor for converting latent to active tuberculosis, and many survivors of a massive nuclear war would have received significant doses of radiation, with important effects on their immune system. Unlike Hiroshima, where immediate assistance from the outside was available, in the post-nuclear world of the 1980s radiation, together with crowding, cold, stress, malnutrition and lack of prior exposure, might predispose the survivors to serious tuberculosis problems.

Plague

Plague has been a known source of epidemics for the past 3,500 years. In the twentieth century alone, over 12 million deaths have been attributed to it.

Plague is endemic among wild rodents in the 11 westernmost states [32, p. 36]. Over 30 types of wild rodents and rabbits have been found infected [33, pp. 45-46]. Cats and dogs can also be infected both experimentally and naturally, or when they ingest infected rodents [34, 35]. Human contact with wild rodents is almost exclusively the source of plague cases in the United States [36].

A nuclear attack would create almost ideal conditions for breaching the "thin protective wall" against plague [33, p. 41]. Large areas of the western United States now relatively devoid of inhabitants may receive an influx of refugees from threatened or devastated urban areas. Relocation plans call for enormous increases in the population of many remote regions [1, p. 35]. Humboldt County, California, for example, would experience a fivefold or more increase in population [37, pp. 958-59]. Millions of urban refugees, unable to obtain shelter in existing dwellings, will build earth-covered "expedient" shelters in undeveloped areas. Such shelters might provide good fallout protection, but they would surely create ideal conditions for transmission of plague from rodents.

Rodents are relatively resistant to plague, develop chronic infections, and may act as a reservoir for the disease. Radiation would increase their susceptibility, as well as that of man. High mortality among wild rodents would then help spread the disease to nearby humans [37, p. 45a]. As the rodents die, the fleas will leave them and search for another host, either animal or man.

Once radiation has subsided in leveled cities, many will head back to reclaim whatever possessions may be found and to reunite with family members [38]. But over 90 percent of the housing will be destroyed and crowding in surviving buildings may be anticipated [39, p. 23].

Conditions in the damaged cities will continue to be favorable for the spread and propagation of plague. The rat population will increase because harborage and food for rats will be available. (A growth rate of 3 to 11 percent per week is expected in the commensal rat population [39, p. 48].) If domestic rats become infected, they will spread the disease among humans. There are two important domestic rat species—*Rattus norvegicus*, found throughout the United States, and *Rattus rattus*, restricted mainly to the Southeast and Pacific Coast [33, pp. 7-8]. The latter is the more efficient vector of plague. Its flea, *Xenopsylla cheopis,* "is the vector par excellence" [33, p. 451]. Its range corresponds with the "at risk" population of the Pacific Coast.

A major danger comes not alone from bubonic plague transmitted by domestic rats, but from man-to-man pneumonic plague [33, p. 43]. Under post-attack conditions, radiation and stress will raise the conversion rate of bubonic to pneumonic plague to 25 percent [11, p. 43].

Pneumonic plague, a highly contagious disease, would be especially dangerous among the survivors of a nuclear attack, crowded together in the few remaining buildings [40]: "Pneumonic plague can spread through the human population at an alarming rate, especially under conditions of crowding such as may be envisioned in the post-attack environment" [41].

The complexity of the control process in the post-attack environment is readily understood from a consideration of the measures required, as noted by Mitchell [20, pp. 208-9]:

- Maintenance of external quarantine
- Maintenance of plague surveillance
- Maintenance of central regional diagnostic laboratory facilities
- Maintenance of antimicrobial drug production
- Maintenance of production of immunologicals
- Capacity for water and sewage control
- Maintenance of food quality control
- Capacity for insect vector control
- Capacity for rodent control

Assuming a uniform risk across the country, it is possible that 6.3 percent of survivors may contract plague; in half it will be fatal [32, pp. 401-3]. Some believe that plague may spread even more widely and probably represents "the major national threat among the set of vectorborne diseases" [32, p. 46].

In the aggregate, deaths from communicable diseases among the survivors may approach 20 to 25 percent [42, pp. 3-10]. Estimates of both the incidence and the mortality of infection in the post-attack world vary widely for different diseases; available figures have been arranged in composite form in Figure 2.

In a computer simulation of the effects of a single nuclear explosion nine miles south of New Orleans, it was calculated that 35 percent of the survivors died from infectious diseases in the first post-attack year, in the absence of medical countermeasures [42, pp. 4-10]. This high mortality from communicable diseases may be compared to the expected deaths from *non*-communicable diseases such as heart disease, diabetes, etc. These are estimated to be between 2.5 to 3 percent of the survivors [42, pp. 3-10]. These may also be compared to excess cancer mortality, estimated at a few percent or less of the survivors [3, p. 156].

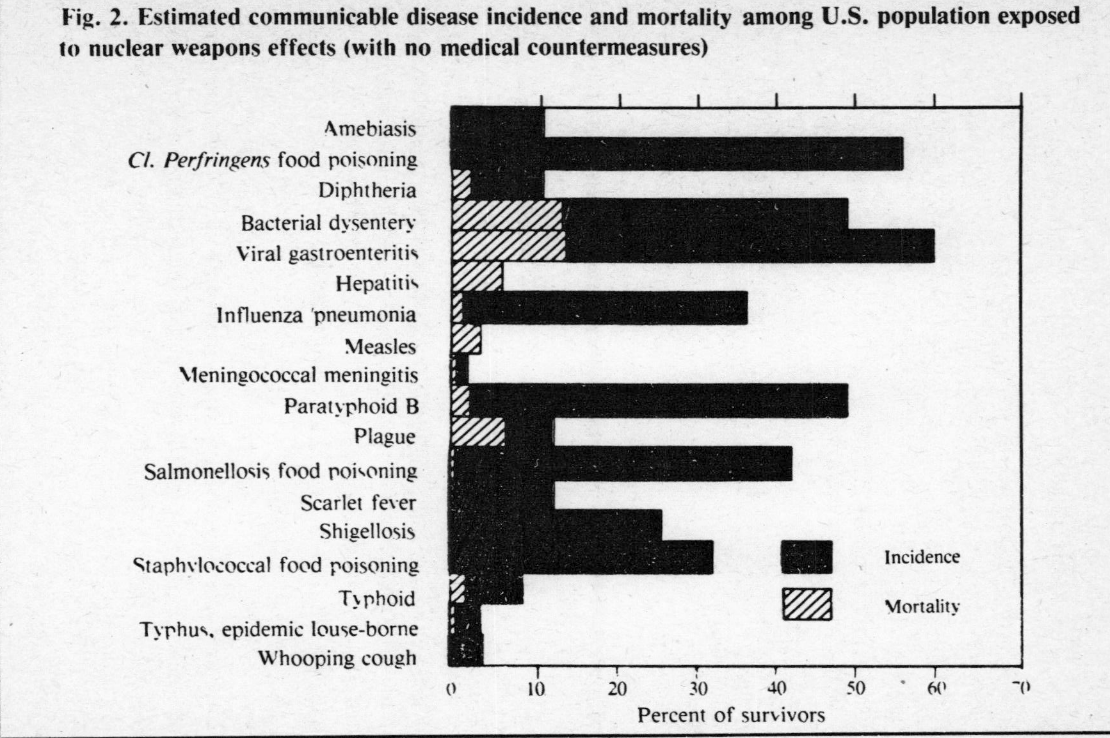

Fig. 2. Estimated communicable disease incidence and mortality among U.S. population exposed to nuclear weapons effects (with no medical countermeasures)

Conclusions

There are numerous factors that point to an increased risk of serious epidemics in the post-attack environment. These factors include the effects of irradiation, malnutrition, and exposure on susceptibility; of unsanitary conditions, lengthy shelter stays, and insect population growth on the transmittal of disease; and of depleted antibiotic stocks, physician shortages, laboratory destruction, and post-attack disorganization on the effectiveness of countermeasures.

Although this threat has been considered small in comparison to the direct effects of nuclear weapons, it requires careful consideration not only of its potential impact on survivors but also on the recovery process itself. Previous studies may have been overly optimistic in their assumptions and analysis:

- The synergistic effects of increased susceptibility, easier disease transmittal, and less effective countermeasures are uncertain.
- Some studies have assumed that the post-attack survival situation is favorable, food is plentiful, and governmental and health organizations are functioning.
- The profound effects of epidemics on post-attack recovery—leading to further famine and disease—have not been adequately calculated.
- Lethal, highly infectious, and largely uncontrollable diseases of non- or low-incidence in the United States are assumed to stay at low-incidence in the post-attack period. A breakout of one or more could greatly increase fatalities.

While no existing data *prove* that "catastrophic" epidemics will occur in the post-attack period, the matter is one of overwhelming importance and uncertainty. What is *certain* is that infection will pose a substantial threat to health and recovery for all those injured by blast, heat, and radiation, and that the resources to grapple with this threat will be inadequate.

1. C.M. Haaland, C.V. Chester and E.P. Wigner, *Survival of the Relocated Population of the U.S. After a Nuclear Attack,* ORNL-5041, Oak Ridge National Laboratory (June 1976), sponsored by Defense Civil Preparedness Agency, pp. 20-21.

2. R.J. Sullivan, W.M. Heller and E.C. Aldridge, Jr., *Candidate U.S. Civil Defense Programs,* SPC-342, System Planning Corporation (March 1978), sponsored by Defense Civil Preparedness Agency, p. 19.

3. R.J. Sullivan, K. Guthe, W.H. Thoms, and F.L. Adelman, *Survival During the First Year After a Nuclear Attack,* SPC-488, System Planning Corporation (Dec. 1979), sponsored by Federal Emergency Management Agency, p. 140.

4. A.W. Voors, *Epidemiological Considerations for the Prevention of Post Nuclear Attack Epidemics,* RM-OU-332-4, Research Triangle Institute (Oct. 1967), sponsored by Office of Civil Defense.

5. Samual Glasstone and Philip J. Dolan, editors, *The Effects of Nuclear Weapons* (Washington, D.C.: U.S. Department of Defense and U.S. Department of Energy, 1977), p. 586.

6. U.S. Senate, *Short- and Long-Term Health Effects on the Surviving Population of a Nuclear War,* Hearing, 96th Cong. (Washington, D.C.: GPO, 1980), p. 62.

7. R.L. Goen, S.L. Brown, E.D. Clark, A.C. Kamradt, H. Lee, R.C. Morey, W.L. Owen, J.W. Ryan and O.S. Yu, *Analysis of National Entity Survival,* SRI 7979-007, Stanford Research Institute (Nov. 1967), sponsored by Office of Civil Defense.

8. Defense Civil Preparedness Agency, *Research Report on Recovery From Nuclear Attack* (Washington, D.C.: DCPA, 1979), p. 12.

9. Cresson H. Kearny, *Nuclear War Survival Skills,* ORNL-5037 (Oak Ridge, Tenn.: Oak Ridge National Laboratory, 1979), p. 79.

10. Robert J. Galiano, *Sustenance of Survivors,* DSA Log No. 6, Decision-Science Applications, Inc. (Dec. 1977), sponsored by Arms Control and Disarmament Agency, sec. 3.4.

11. William M. Brown and Doris Yokelson, *Postattack Recovery Strategies,* HI-3100-RR, Hudson Institute, Inc. (Nov. 1980), sponsored by Federal Emergency Management Agency, p. 30.

12. Robert J. Galiano, *Medical and Health Problems,* DSA Report No. 4, Decision-Science Applications, Inc. (Aug. 1977), sponsored by Arms Control and Disarmament Agency, sec. 1.4.

13. R.U. Ayres, Environmental Effects of Nuclear Weapons (Groton on Hudson, NY, The Hudson Institute, 1965).

14. Arms Control and Disarmament Agency, *Effects of Nuclear War* (Washington, D.C.: ACDA, 1979), p. 23.

15. "Biological Warfare," *Encyclopedia Britannica,* 1965 ed. pp. 641-2.

16. William J. Broad, "Nuclear Pulse (I): Awakening to the Chaos Factor," *Science* (May 29, 1981).

17. D.R. Johnston, M.N. Laney, R.L. Chessin, and D.G. Warren, *Study of Crisis Administration of Hospital Patients; and Study of Management of Medical Problems Resulting from Population Relocation,* RTI/1532/00-04F, Research Triangle Institute (Sept. 1978), prepared for Defense Civil Preparedness Agency, p. V-32.

18. M. Staackmann, W.H. Van Horn, and C.R. Foget, *Damage to the Drug Industry From Nuclear Attack and Resulting Requirements For Repair and Reclamation,* URS 796-4, URS Research Company (July 1970), sponsored by Office of Civil Defense, p. 3.

19. U.S. Congress, Joint Committee on Defense Production, *Economic and Social Consequences of Nuclear Attacks on the United States,* prepared by Arthur Katz, published by the Senate Committee on Banking, Housing and Urban Affairs (Washington, D.C.: GPO, 1979), pp. 19-20.

20. H.H. Mitchell, *Guidelines for the Control of Communicable Disease in the Postattack Environment,* RDA-TR-051-DCPA, R & D Associates (July 1972), sponsored by Defense Civil Preparedness Agency, pp. 205-6.

21. R.U. Ayres, *Environmental Effects of Nuclear Weapons,* HI-518-RR Hudson Institute (1965), sponsored by Office of Civil Defense.

22. H.H. Mitchell, *Survey of the Infectious Disease Problem as it Relates to the Post-Attack Environment,* RM-5090-TAB, The RAND Corporation, (Aug. 1966), sponsored by U.S. Atomic Energy Commission, p. 50.

23. "Death Rates from Selected Causes," *Information Please Almanac* (New York: Simon and Schuster, 1980), pp. 407-8, 811.

24. Center for Disease Control, *Tuberculosis in the United States 1978* (Washington, D.C.: U.S. Department of Health and Human Services, 1980), p. 2.

25. Alan Leff, William Lester, and Whitney W. Addington, "Tuberculosis: A Chemotherapeutic Triumph but a Persistent Socioeconomic Problem," *Archives of Internal Medicine,* 139 (Dec. 1979), p. 1376.

26. H.H. Mitchell, *The Problem of Tuberculosis in the Postattack Environment,* RM-5362-PR, The RAND Corporation (June 1967), sponsored by United States Air Force, pp. 10-14.

27. Charles W. Ochs, "The Epidemiology of Tuberculosis," *Journal of the American Medical Association,* 179 (Jan. 27, 1962), pp. 87-92.

28. N.S. Scrimshaw, C.E. Taylor, and J.E. Gordon, *Interactions of Nutrition and Infection,* WHO Monograph No. 57 (Geneva: World Health Organization, 1968), pp. 61, 69-70, 88, 98.

29. "Tuberculosis and Nutrition," *Tubercle* (Jan. 1948), pp. 20-21.

30. R.S. Pogrund, *Nutrition in the Postattack Environment,* RM-5052-TAB, The RAND Corporation (Dec. 1966), prepared for U.S. Atomic Energy Commission.

31. Raymon W. Turner, and Dorothy R. Hollingsworth, "Tuberculosis in Hiroshima," *Yale Journal of Biology and Medicine* (Oct. 1963), pp. 180-81.

32. T. Johnson and D.R. Johnston, *Vectorborne Disease and Control,* R-OU-303, Research Triangle Institute (June 1968), sponsored by Office of Civil Defense, p. 36.

33. H.H. Mitchell, *Plague in the United States: An Assessment of Its Significance As a Problem Following a Thermonuclear War,* RM-4868-TAB, The RAND Corporation (June 1966), sponsored by U.S. Atomic Energy Commission, pp. 45-46.

34. J.H. Rust, Jr., D.C. Cavanaugh, R. O'Shita, and J.D. Marshall, Jr., "The Role of Domestic Animals in the Epidemiology of Plague. I. Experimental Infection of Dogs and Cats," *The Journal of Infectious Diseases,* 124 (Nov. 1971), pp. 522-26.

35. J.H. Rust, Jr., B.E. Miller, M. Bahmanyar, J.D. Marshall, Jr., S. Purnaveja, D.C. Cavanaugh, and U.S. Tin Hla, "The Role of Domestic Animals in the

Epidemiology of Plague. II. Antibody to *Yersinaia pestis* in Sera of Dogs and Cats," *The Journal of Infectious Diseases,* 124 (Nov. 1971), pp. 527-31.

36. Abram S. Benenson, "Plague," in *Communicable and Infectious Diseases,* F.H. Top, Sr., and P.F.Wehrle, eds. (St. Louis: The C.V. Mosby Company, 1972), p. 451.

37. "Census and Areas of Counties and States," *The World Almanac & Book of Facts 1981* (New York: Newspaper Enterprises Association, Inc., 1980), pp. 958-59.

38. Fred Charles Iklé, *The Social Impact of Bomb Destruction* (Norman, Oklahoma: University of Oklahoma Press, 1958), pp. 211-24.

39. Arms Control and Disarmament Agency, *Effects of Nuclear War* (Washington, D.C.: GPO, 1979), p. 23.

40. M. Bahmanyar and D.C. Cavanaugh, *Plague Manual* (Geneva: World Health Organization, 1976), p. 59.

41. H.H. Mitchell, *Guidelines for the Control of Communicable Diseases in the Postattack Environment,* RDA-TR-051-DCPA, R & D Associates (July 1972), sponsored by Defense Civil Preparedness Agency, p. 103.

42. E.L. Hill, A.W. Voors, R.O. Lyday, Jr., J.N. Pyecha, J.B. Hallan, J.T. Ryan, and C.N. Dillard, *National Emergency Health Preparedness Study, Including the Development and Testing of a Total Emergency Health Care System Model,* R-OU-332, Research Triangle Institute (Nov. 1968), prepared for Office of Civil Defense, Office of the Secretary of the Army, pp. 3-10.

43. D.R. Johnston, M.E. Fogel, A.W. Voors, and E.L. Hill, *Postattack Prevention and Control of Enteric Diseases,* Research Triangle Institute (Sept. 1969), prepared for Office of Civil Defense, Office of the Secretary of Defense.

44. R.O. Lyday, Jr., J.N. Pyecha, and E.L. Hill, *Postattack Medical Care Impact on Survivors' Work Force,* Research Triangle Institute (July 1976), prepared for Defense Civil Preparedness Agency.

45. R.S. Titchen, *Late Postnuclear Attack Health Problems Study,* presented at the 30th National Meeting Operations Research Society of America, 1968.

46. A.W. Voors and B.S.H. Harris, *Postattack Communicable Respiratory Diseases,* R-OU-493, Research Triangle Institute (Nov. 1970), prepared for Office of Civil Defense, Office of the Secretary of the Army.

V TREATMENT

"One medical strategy is still available—prevention."

—H. Jack Geiger

What is the best preparation for post-attack recovery?
Pre-attack restraint.

16 Preventing nuclear war

ROGER FISHER

"Preventing Nuclear War"? "Boy, have *you* got a problem." That reaction a few minutes ago to the title of these remarks reminded me of an incident when during World War II, I was a B-17 weather reconnaisance officer. One fine day we were in Newfoundland test-flying a new engine that replaced one we had lost. Our pilot's rank was only that of flight officer because he had been court martialed so frequently for his wild activities; but he was highly skillful.

He took us up to about 14,000 feet and then, to give the new engine a rigorous test, he stopped the other three and feathered their propellers into the wind. It is rather impressive to see what a B-17 can do on one engine. But then, just for a lark, the pilot feathered the fourth propeller and turned off that fourth engine. With all four propellers stationary, we glided, somewhat like a stone, toward the rocks and forests of Newfoundland.

After a minute or so the pilot pushed the button to unfeather. Only then did he remember: In order to unfeather the propeller you had to have electric power, and in order to have electric power you had to have at least one engine going. As we were buckling on our parachutes, the co-pilot burst out laughing. Turning to the pilot he said, "Boy, have *you* got a problem!"

As with the crew of that B-17, we're all in this together. People tend, however, to put the problem of preventing nuclear war on somebody else's agenda. But whoever is responsible for creating the danger, we're all on board one fragile spacecraft. The risk is high. What can we do to reduce it?

There are two kinds of reasons for the high risk: hardware

reasons and people reasons: We—and the military—tend to focus on the hardware: nuclear explosives and the means for their delivery. We think about the terrible numbers of terrible weapons. We count them by the hundreds, by the thousands and by the tens of thousands. There are clearly too many. There are too many fingers on the trigger. There are too many hands through which weapons pass in Europe, in the United States and in the Soviet Union.

Yes, changes should be made in the hardware. I believe we should stop all nuclear weapons production; we should cut back on our stockpiles. But even if we should succeed in stopping production, and even if we should succeed in bringing about significant reductions, there will still be thousands of nuclear weapons. We keep our attention on the hardware. The military think it is the answer; we think it is the problem. In my judgment it is not the most serious problem.

The U.S. Air Force and the U.S. Navy both have enough weapons to blow each other up; and they have disagreements. There are serious disputes between the Air Force and the Navy: disputes that mean jobs, careers; disputes that are sometimes more serious in practical consequences than those between the United States and the Soviet Union. But the two services have learned to fight out their differences before the Senate Appropriations Committee, before the Secretary of Defense, in the White House and on the football field.

The case of the Navy and the Air Force demonstrates, in a crude way, that the problem is not just in the hardware; it is in our heads. It lies in the way we think about nuclear weapons. And if the problem lies in the way we think, then that's where the answer lies. In Pogo's immortal phrase, "We have met the enemy and they are us."

The danger of nuclear war is so great primarily because of the mental box we put ourselves in. We all have working assumptions that remain unexamined. It is these assumptions that make the world so dangerous. Let me suggest three sets of mistaken assumptions about: (1) our goals, that is, the ends we are trying to pursue; (2) the means for pursuing those ends; and (3) whose job it is to do what.

• First, what are our goals? Internationally, we think we want to "win." We go back to primitive notions of victory. "Who's winning?" But that is an area in which we must change our thinking. Internationally and domestically, we do not really want a system in which any one side—even our own—wins all the time. Yet this concept of "winning"—that there is such a thing and that it is our dominant objective—is one of our fundamental beliefs.

In fact, like a poker player, we have three kinds of objectives. One is to win the hand. Whatever it is we think we want, we want it now. We want victory. The second is to be in a good position for future hands. We want a reputation and chips on the table so that we can influence future events. In other words, we want power. Our third objective is not to have the table kicked over, the house burned down, or our opponent pull a gun. We want peace.

We want victory; we want power; we want peace. Exploding nuclear weapons will not help us achieve any one of them. We have to re-examine rigorously our working assumption that in a future war we would want to "win." What do we mean by "win?" What would our purpose be?

Last year I gave the officers of the NATO Defense College in Rome a hypothetical war in Europe and asked them to work out NATO's war aims. The "war" was presumed to have grown out of a general strike in East Germany, with Soviet and West German tanks fighting on both sides of the border. Deterrence had failed. I told the officers: "You are in charge of the hotline message to Moscow. What is the purpose of this war? What are you trying to do?" At first they thought they knew—win! Very simple. But what did that mean? What was the purpose of the war? They began to realize that NATO did not plan to conquer the Soviet Union acre by acre as the Allies had conquered Germany in World War II. They did not plan physically to impose their will on the Soviet Union. They were seeking a Soviet decision. That was the only way they could have a successful outcome.

With further thought they reached a second conclusion: they were not going to ask for unconditional surrender. That gave them a specific task: Just define the Soviet decision that would constitute success for NATO and that NATO could reasonably expect the Soviet Union to make. The officers worked through the day considering how the Russians probably saw their choice, how we wanted them to see it, and what kind of "victory" for us we could realistically expect the Soviet Union to agree to.

It turned out that the only plausible objective was to stop the war. "Winning" meant ending the war on acceptable terms. The goal was some kind of cease-fire, the sooner the better. It was with difficulty and even pain that some officers discovered that winning meant stopping, even if some Soviet troops remained in West Germany; even with only a promise to restore the *status quo ante.*

They found it hard to draft a fair cease-fire that didn't sound like a unilateral Western ultimatum. It might say, "Stop firing at 0100 hours tomorrow, promise to withdraw, promise to restore the status quo within 48 hours, and we will meet in Vienna to talk about serious problems as soon as the status quo is 'more or less' restored." But the NATO draftsmen did not know whether the Soviets would prefer Geneva to Vienna or whether they wanted 0200 hours instead of 0100 hours, etc.

Someone creatively suggested, "Wouldn't it be a good idea if right now we worked out with the Russians some standby cease-fire terms? Then in a crisis we wouldn't have to be demanding that they accept our terms or they demanding that we accept theirs. Let's produce some cease-fire drafts that we can both accept." One of the other officers was incredulous: "What did you say?

You are going to negotiate the armistice before the war begins? In that case, why have the war?"

The need to re-examine assumptions about our foreign policy objectives is also demonstrated by our self-centered definition of national security. Typically, political leaders and journalists alike suggest that the primary goal of foreign policy is national security, and only after that has been assured should we worry about our relations with the Soviet Union.

Such thinking assumes that we can be secure while the Soviet Union is insecure—that somehow we can be safe while the Soviet Union faces a high risk of nuclear war. But in any nuclear war between the United States and the Soviet Union missiles will go both ways. There is no way we can make the world more dangerous for them without also making it more dangerous for ourselves. The less secure the Soviets feel, the more they will be doing about it, and the less secure we will become. Security is a joint problem.

We must make the Soviet Union share the responsibility for our security problem. We should say, "Look, you Russians have to understand why we build these missiles and how it looks to us when you behave as you do. You must take some responsibility for helping us deal with our security problem." Similarly, we must take on responsibility for dealing with their security problem. We cannot make our end of the boat safer by making the Soviet end more likely to capsize. We cannot improve our security by making nuclear war more likely for them. We can't "win" security from nuclear war unless they win it too. Any contrary assumption is dangerous.

Here I may point out that we in the peace movement do not always practice what we preach. I am always ready to tell friends at the Pentagon that it does no good to call Soviet officials idiots, but am likely to add, "Don't you see that, you idiot"? We who are concerned with reducing the risks of war often think that our job is to "win" the war against hawks. In advancing our interests we assume that our adversaries have none worth considering. But our task is not to win a battle. Instead, we have to find out what the other side's legitimate concerns are, and we have to help solve their legitimate problems in order to solve our own. At every level, domestically and internationally, we need to re-examine our working assumptions. We are not seeking to win a war, but to gain a peace.

• A second set of dangerous assumptions are those we make about how to pursue our objectives. The basic mistaken assumption is that for every problem there is a military solution. We will first try diplomacy. We will talk about it; we will negotiate. But if that doesn't solve the problem we assume that we can always resort to force. We tend to assume that if we have the will and the courage, and are prepared to pay the price, then we can always solve the problem by military means. Wrong. For the world's big problems there is no military solution. Nuclear war is not a solution. It is worse than any problem it might "solve."

We have mislearned from the past. During World War II the Allies could physically impose a result on Hitler and his country. Acre by acre it was done. But the world has changed. We can no longer impose such a result on any nuclear power. We cannot physically make things happen. The only means we have available is to try to change someone's mind.

There is no way in which nuclear hardware can bring about a physical solution to any problem we face except the population problem. Just as you cannot make a marriage work by dynamite or make a town work by blowing it up, there is no way we can make the world work by using nuclear bombs. When people hear that they say, "Yes, that's true." Yet they go right ahead, operating on the assumption that there are military solutions.

Like Linus in the Charlie Brown comic strip, we cling to our security blanket, military hardware. Both U.S. and Soviet officials clutch their plutonium security blankets as though somehow they offer real security. Somehow, we think, this bomb, this hardware, will give us strength, will protect us. We will be able to avoid the necessity of dealing with the real world. Our assumption is that the problem is simple—it's us against them. We want to believe in a quick fix. It is like cowboys and Indians. Whatever the problem, John Wayne will arrive with his six-guns blazing and the problem will be solved.

Those are our common assumptions about how to deal with international problems. We operate on them although most of us know they are not true. The fact that conventional weapons remain useful and that conventional wars continue reinforces our mistaken assumptions about the use of nuclear weapons.

We have far better ways to deal with international problems. Break up big problems into manageable pieces. Look at each item on its merits. Sit down side-by-side and discuss it. Don't concentrate on what our adversaries say their positions are, but try to understand and deal with their interests. Communicate and listen. What's in their minds? What's bothering them? How would we feel?

If you were sitting in Moscow and looking off to the left saw Japan thinking of rearming; if you saw your long-time strongest ally, China, with a 4,000-mile common frontier now your worst enemy; if you saw Pakistan apparently getting a nuclear bomb; if you heard Western voices saying, "We must help the rebels in Afghanistan"; if you saw American military equipment now in the Gulf, in Saudi Arabia, in Egypt and in Israel; if you saw Greece rejoining NATO and Turkey in the hands of a military government; and if cruise missiles were about to be located in West Germany—how must that all look from Moscow? We should put ourselves in their shoes and understand their problems. The only way we can succeed is to affect their future thinking. The starting point is to understand their present thinking.

Second, we have to invent wise solutions. We have to figure out not just good arguments, but good ways to reconcile our differing interests. And they must participate in that process. There is no way, in any conflict, in which one side can produce the right answer. The understanding that comes from both sides working on a problem, and the acceptability that comes from joint participation in a solution, make any good answer better. We need to engage in joint problem-solving.

That same process of working together is equally applicable to our domestic differences. The peace movement is not the only source of wisdom; we are part of the conflict. There are a lot of people in this country who have legitimate concerns about the Soviet Union. We must try to understand these concerns and meet them, not carry on a war. We need to put ourselves in their shoes —in Pentagon shoes. We have to listen as well as talk. With their participation, we will invent better solutions.

By this process, we will promote joint learning, not just at the intellectual level but at the level of feeling, of emotion, of caring, the level of concern. International conflict is too often dealt with cerebrally, too often dealt with as a hypothetical problem out there. We need not only to apply what we know, but to keep on learning about human behavior, how to affect our own behavior and that of others, not just manipulate it.

The danger of nuclear war lies largely within us. It lies in how we think about winning, in how we define success, and in our illusions of being able to impose results.

• The danger also comes from my third set of assumptions—about whose job it is to reduce the risk of war. If there were a military solution, there would be a case for leaving it to the military—to policy-science experts, and to professional strategists. Physicians, for example, have said: "We are just concerned with the medical aspects of nuclear war and will limit ourselves to that area. We will tell you how bad a nuclear war would be. It is somebody else's job to prevent it."

Such statements rest on the assumption that the solution is in the hardware department. But we are not facing a technical military problem "out there." The solution lies right here: in changing our own assumptions and those of other people; in growing up; in abandoning our plutonium security blanket.

The Soviet Union and the United States cling to nuclear weapons as symbols of security; other national leaders want them. If someone is clinging to a plutonium blanket which is bad for his health, you do not call in an engineer and say, "Design a better plutonium blanket." The problem is in the heads of those who are clinging to it.

There is no one I know who has a professional license in the skills of reducing the risk of nuclear war. Fortunately, however, no professional license is required. But who are those with skills in dealing with psychological problems like denial, like distancing, like turning flesh and blood issues into abstract problems through the use of jargon? Who is likely to notice that people are denying responsiblity because a problem seems too overwhelming? Nuclear engineers? I think not.

Now back to the B-17 over the hills of Newfoundland. The copilot was saying to the pilot: "Boy, have *you* got a problem." Well, we didn't crash; we weren't all killed. On that plane we had a buck sergeant who remembered that back behind the bomb bay we had a putt-putt generator for use in case we had to land at some emergency air field that did not have any electric power to start the engines. The sergeant found it. He fiddled with the carburetor; wrapped a rope around the flywheel a few times; pulled it and pulled it; got the generator going and before we were down to 3,000 feet we had electricity. The pilot restarted the engines, and we were all safe. Now saving that plane was not the sergeant's job in the sense that he created the risk. The danger we were in was not his fault or his responsibility. But it was his job in the sense that he had an opportunity to do something about it.

We professionals tend to define our roles narrowly. I sometimes ask my law students: "What would have been the responsibility of a professional musician judging Nero's performance on the fiddle while Rome burned? Should he limit himself to discussing the music?" A member of the lay public would probably get a bucket and put out the fire. By becoming professionals do we become less responsible? Can we say, "No, I'm a professional. I'm not a firefighter. That's someone else's job."?

Such special knowledge and training as we have may not make it obligatory for us to try to prevent nuclear war. Rather, it gives us an opportunity. My notion of whose job something is is best defined by who has an opportunity. We have an opportunity. I encourage you, as I encourage myself, to use it. The world is at risk. The very danger of nuclear war means that there is more opportunity to make a difference than ever before.

If everyone with any significant power made the right decision every time, that's as near utopia as we can get. There are only three reasons they don't. One is that they are poor decision-makers. Our job is to change them; that is what politics is about. Second, they are operating on bad assumptions, thinking poorly. Our job is to correct their assumptions. And the third possibility is that they are subject to harmful constraints. Our job is to free them from those constraints.

In a simple chart I have put all problems on the left-hand side, divided into those three parts. Across the top are the activities in which we can engage: research on facts and theory; communication in terms of learning ourselves and teaching others; devising things to do (that is, turning a problem into a possible answer, inventing possible proposals or action ideas), getting a proposal onto somebody's agenda; advocating ideas; or doing something ourselves.

To get a wise decision we need good answers *in every box.* No amount of useful research will overcome poor deciders; no number of good deciders will overcome bad assumptions or harmful constraints. Somebody has to invent what's to be done. Somebody has to persuade others that it is a good idea and somebody

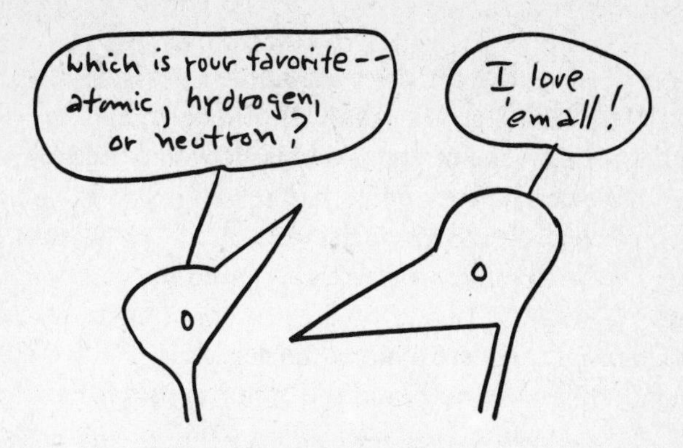

Chart of useful activities

Avoiding nuclear war will require wise decision-makers, who make wise assumptions on ends and means, and who are free from harmful constraints. Which part of that problem do you want to work on? Which activity will best harness your interests and abilities? To get wise results someone must devote attention to each box in the matrix. To get a wise decision answers are needed in every box.

PROBLEMS	ACTIVITIES					
	Research: facts, theory	Communicate: learn, teach	Devise things to do	Build an agenda	Advocate	Do it yourself
Poor deciders						
Poor thinking on ends on means						
Harmful constraints						

has to do it. All those activities are needed for each category of problem. There is enough to keep all of us busy.

All of us can do any of those things. No single activity will be sufficient. We need theory on how to reduce instability. We need to develop knowledge about nuclear war, about the consequences and about ways to reduce the risks. We need to communicate the knowledge both to the public, who constrain our decision-makers, and to the people who are making the decisions. We need to communicate both the bad news and the opportunities for reducing it.

If all we do is deliver bad news and say that there's nothing we can do about it, the bad news does not become operational. We have to turn that news into something we can do.

My favorite activity is inventing. An early arms control proposal dealt with the problem of distancing that the President would have in the circumstances of facing a decision about nuclear war. There is a young man, probably a Navy officer, who accompanies the President. This young man has a black attache case which contains the codes that are needed to fire nuclear weapons. I could see the President at a staff meeting considering nuclear war as an abstract question. He might conclude: "On SIOP Plan One, the decision is affirmative. Communicate the Alpha line XYZ." Such jargon holds what is involved at a distance.

My suggestion was quite simple: Put that needed code number in a little capsule, and then implant that capsule right next to the heart of a volunteer. The volunteer would carry with him a big, heavy butcher knife as he accompanied the President. If ever the President wanted to fire nuclear weapons, the only way he could do so would be for him first, with his own hands, to kill one human being. The President says, "George, I'm sorry but tens of millions must die." He has to look at someone and realize what death is—what an innocent death is. Blood on the White House carpet. It's reality brought home.

When I suggested this to friends in the Pentagon they said, "My God, that's terrible. Having to kill someone would distort the

President's judgment. He might never push the button."

Whether or not that particular idea has any merit, there is lots to do. Action is required to convince the public that it is in our interest to have the Soviets feel secure rather than insecure. Much of the press apparently thinks that the more terrified the Soviets are the more we benefit. The Committee on the Present Danger (perhaps better called the Committee on Increasing the Present Danger) ignores the fact that if we raise the risk of nuclear war for the Soviet Union we also raise it for ourselves.

If you don't know what to do, that's great. That gives you something to do right there. Get some friends together on Saturday morning and generate some ideas. Separate this inventing process from the later process of deciding among them. Identify three or four other people who might make a decision of some significance. What can you do to increase the chance they'll make some desired decision next week? Whoever it is—journalists, congressmen, governors, legislators, newspaper editors, businessmen, a civic organization, a medical association, a friend of President Reagan's, a school teacher, a publisher—what are some things they might do that would illuminate our faulty working assumptions to help establish better ones? Figuring out what to do is itself an excellent activity. In intellectual efforts, as in gunnery, aiming is crucial.

Don't wait to be instructed. Take charge. This is not an organized campaign that someone else is going to run. If you share these concerns, get involved. There is a lot to do to reduce the risk of nuclear war. Reading, writing, talking, perhaps a radio program, or perhaps a letter-writing campaign to your congressman. Tell him, "Grow up. Give up your plutonium security blanket."

But perhaps you are still holding on to your own security blanket, that neat definition of your job. The security blanket most of us cling to is, "Don't blame me. It's not my job to plan nuclear strategy. I'm not responsible for the risk of nuclear war." You can give up that security blanket any time.

The way to enlist support is not to burden others with guilt but to provide them with an opportunity to volunteer. I find it an exciting venture. It is a glorious world outside. There are people to be loved and pleasures to share. We should not let details of past wars and the threat of the future take away the fun and the joy we can have working together on a challenging task. I see no reason to be gloomy about trying to save the world. There is more exhilaration, more challenge, more zest in tilting at windmills than in any routine job. Be involved, not just intellectually but emotionally. Here is a chance to work together with affection, with caring, with feeling. Feel some of your emotions. Don't be uptight. You don't have to be simply a doctor, a lawyer, or a merchant. We are human beings. Be human.

People have struggled all of their lives to clear ten acres of ground or simply to maintain themselves and their family. Look at the opportunity we have. Few people in history have been given such a chance—a chance to apply our convictions, our values, our highest moral goals with such competence as our professional skills may give us. A chance to work with others—to have the satisfaction that comes from playing a role, however small, in a constructive enterprise. It's not compulsory. So much the better. But what challenge could be greater? We have an opportunity to improve the chance of human survival.

In medicine there is a traditional call that strikes a nice balance between duty and opportunity, that invites us to lend a hand with all the skill and compassion we can muster: "Is there a doctor in the house?"

17 The physician's commitment

BERNARD LOWN

The baton in the relay race against a nuclear Armageddon is being passed from physicist to physician. The scientists, who in a Faustian bargain first released the atomic genie, have been alerting the public for the past 36 years to the mortal danger facing humankind. The authoritative voices and dire predictions of physicists have not impeded the spiralling cumulation of overkill in its escalation to an inexorable nuclear confrontation.

Physicians do not possess special knowledge of direct experience in atomic matters. They do, however, have unique expertise in areas relating to the medical consequences of nuclear war, to the possibilities of medical care in a post-attack period, to the noninvolvement and denial by the intended victims, to the malfunctioning of technology and to the aberration of personality which may trigger a nuclear exchange. The physicians' analysis is precise, clinical and exorcises the mystifying verbiage, *Manichean* oversimplifications and sanitized statistics of the strategic experts. The physicians' movement is compelled by a growing conviction that nuclear war is the number one health threat and perhaps constitutes the final epidemic for which the only remedy is prevention.

The underlying thrust to the physicians' movement is that of educating colleagues, and through them their endangered patients. The physician, as a health provider and interpreter of complex scientific facts, maintains a position of credibility with the individual being served. As such, the public can trust the physician in expressing an opposition to the nuclear arms race that derives from a deep commitment to preserving human life. The objective of the physicians' movement is to compel society to confront the essential fact that nuclear weapons and human beings cannot coexist for long on this small planet.

Nuclear war is a term of deception. War has been thought of as an extension of politics, having defined objectives, weapons of ascertainable destructiveness, defense measures to limit casualties, and physicians to care for the wounded, winners and losers. But how is this relevant to an aftermath wherein blast, firestorm, and radioactive fallout destroy the very social fabric? What is the meaning of victory in the wake of a holocaust? It is essential to stop perceiving nuclear bombs as weapons, for they are not weapons but instruments of genocide. Nuclear war between the superpowers can only be an incomprehensible act of collective suicide, prefigured in miniature by the Jones group in Guyana.

Will a substantial number of the 3 million doctors worldwide respond to this unprecedented threat to the public health? Will physicians from the Soviet bloc and elsewhere jettison partisan political and national conceptions and respond to this historic challenge? A preliminary answer was provided at the First Congress of the International Physicians for the Prevention of Nuclear War. Leaders in medicine from the United States, the Soviet Union, Japan, and eight other countries attended the First Congress. The discussions and presentations were typical of a medical scientific meeting and were free of political posturing. Upon consideration of the immediate and delayed consequences of an all-out nuclear war between the superpowers, and the physician's role in its aftermath, conference participants concluded: "The earth would be seared; the skies would be heavy with lethal concentrations of radioactive particles, and no response to medical needs should be expected from medicine." (See page 241 for an excerpt from the IPPNW proceedings.)

But the injury to humankind does not await detonation of nuclear devices. Even in the absence of war, the arms race imposes a heavy burden, especially upon those who live in underdeveloped countries. World military expenditures are now exceeding $500 billion annually or $1.4 billion daily. This massive diversion of scarce resources diminishes development of knowledge, technology and manpower that could address global ecologic and overpopulation problems. A small fraction of these expenditures for the military could provide the world with adequate food and water, sanitation, housing, education, modern health care and the like.

Physicians can cite numerous illustrations of the power of modest investments in health. For example, slightly over a decade ago, smallpox was endemic in 33 countries; 1.2 billion people endured 10 to 15 million cases and two million lives were lost annually. The World Health Organization campaign for smallpox eradication was successfully completed in a decade. The last case of naturally occurring smallpox was diagnosed in Somalia in 1977. This signal achievement required an investment of $300 million, or approximately five hours of the cost of military budgets. The lack of clean water accounts for 80 percent of all the world's illness. With a diversion of funds consumed by three weeks of the arms race, the world could obtain a sanitary water supply for all of its inhabitants.

Participants at the IPPNW Congress focused on the psychologic factors contributing to the nuclear arms race. Living with the possibility of imminent annhilation has created a new reality for humanity with profound and widespread psychologic effects. Not only does each person have to deal with the possibility of one's own agony or sudden death, but also with the dissolution of humanity. Membership in a family, community, society and nation provides the psychologic means for coming to terms with individual death. Something of the "I" survives in the social germ plasm providing a symbolic continuity or immortality. But suppose nothing survives! How does one come to terms with such an absolute discontinuity?

The entrapment in helplessness and terror promotes diverse

psychologic defense mechanisms including numbing, denial, entrusting decisions to experts or political leaders. Although denial reduces the level of anxiety, thereby enabling human beings to function, there is an attendant cost, namely, the impossibility of dealing substantively with the denied threat. At the same time there is regression to primitive thinking patterns, with reliance on spurious notions of strength, achieved by increasing the overkill capacity and pervaded by concepts of winning or losing. The climate of fear and distrust leads to perceptual distortion with complex differences between diverse social systems reduced to mortal combat between forces of good and evil. The imagined enemy eventually is extruded from the human family and reduced to an inanimate object whose annihilation is devoid of moral dimension. The dehumanization of the presumed adversary redounds to our own dehumanization.

The IPPNW Congress demonstrated that Eastern and Western physicians concur on the non-survivability of their respective societies in the event of nuclear war between the superpowers. They likewise agreed on the impossibility of limiting such a conflict once begun. Little credence was given to the value of civil defense measures in reducing casualties and enhancing survival. An additional conclusion was that increases in nuclear stockpiles serve to augment world insecurity. The escalation in weapons thus becomes a catalyst maintaining the mad cybernetics of the arms race.

A world movement of physicians is now emerging. Physicians are not unmindful of the enormity of the challenge, but the medical worker is trained to devise practical solutions to seemingly insoluble human problems. The aim is to educate and alert the widest public. The guarded optimism pervading the physicians' movement derives from an abiding faith in the concept that what humanity creates, humanity can control.

18 International Physicians for the Prevention of Nuclear War

It is difficult for us, as physicians, to describe adequately the human suffering that would ensue a nuclear war. Hundreds of thousands would suffer third-degree burns, multiple crushing injuries and fractures, hemorrhage, secondary infection, and combinations of all of these. When we contemplate disasters, we often assume that abundant medical resources and personnel will be available. But contemporary nuclear war would inevitably destroy hospitals and other medical facilities, kill and disable most medical personnel, and prevent surviving physicians from coming to the aid of the injured because of widespread radiation dangers. The hundreds of thousands of burned and otherwise wounded people would not have any medical care as we now conceive of it: no morphine for pain, no intravenous fluids, no emergency surgery, no antibiotics, no dressings, no skilled nursing, and little or no food or water. The survivors will envy the dead.

It is known from the Japanese experience that in the immediate aftermath of an explosion, and for many months thereafter, the survivors suffer not only from their physical injuries—radiation sickness, burns, and other trauma—but also from profound psychological shock caused by their exposure to such overwhelming destruction and mass death. The problem is social as well as individual. The social fabric upon which human existence depends would be irreparably damaged.

Those who did not perish during the initial attack would face serious—even lifelong—dangers. Many exposed persons would be at increased risk, throughout the remainder of their lives, of leukemia and a variety of malignant tumors. The risk is emotional as well as physical. Tens of thousands would live with the fear of developing cancer or of transmitting genetic defects, for they would understand that nuclear weapons, unlike conventional weapons, have memories—long, radioactive memories. Children are known to be particularly susceptible to most of these effects. Exposure of fetuses would result in the birth of children with small head size, mental retardation, and impaired growth and development. Many exposed persons would develop radiation cataracts and chromosomal aberrations.

In one likely and specific scenario that we have considered, an all-out nuclear war between the United States and the Soviet Union in the mid-1980s, it is likely that the population will be devastated:

• over 200,000,000 men, women, and children will be killed immediately;

• over 60,000,000 will be injured.

And among the injured:

• 30,000,000 will experience radiation sickness;

• 20,000,000 will experience trauma and burns; and

• 10,000,000 will experience trauma, burns, and radiation sickness.

Medical resources will be incapable of coping with those injured by blast, thermal energy, and radiation:

• 80 percent of physicians will die;

• 80 percent of hospital beds will be destroyed;

• stores of blood plasma, antibiotics and drugs will be destroyed or severely compromised;

• food and water will be extensively contaminated; and

• transportation and communication facilities will be destroyed.

The disaster will have continuing consequences:

• food production will be profoundly altered;

• fallout will constitute a continuing problem;

• survivors with altered immunity, malnutrition, an unsanitary environment, and severe exposure problems will be subject to lethal enteric infections;

- profound changes in weather caused by particulates and reduction of atmospheric ozone with attendant alterations in man, animal, and plant species may be anticipated;
- among long-term survivors, a striking increase in leukemia and other malignancies will be observed, most severe in those who are children at the time of exposure.

Civil defense will be unable to alter the death and devastation described above to any appreciable extent.

Nuclear war would be the ultimate human and environmental disaster. The immediate and long-term destruction of human life and health would be on an unprecedented scale, threatening the very survival of civilization.

The threat of its occurrence is at a dangerous level and is steadily increasing. But even in the absence of nuclear war, invaluable and limited resources are being diverted unproductively to the nuclear arms race, leaving essential human, social, medical, and economic needs unmet.

For these reasons, physicians in all countries must work toward the prevention of nuclear war and for the elimination of all nuclear weapons. Physicians can play a particularly effective role because they are dedicated to the prevention of illness, care of the sick and protection of human life; they have special knowledge of the problems of medical response in nuclear war; they can work together with their colleagues without regard to national boundaries; and they are educators who have the opportunity to inform themselves, their colleagues in the health professions, and the general public.

What physicians can do to prevent nuclear war:

- Review available information on the medical implications of nuclear weapons, nuclear war and related subjects.
- Provide information by lectures, publications and other

means to the medical and related professions and to the public on the subject of nuclear war.

- Bring to the attention of all concerned with public policy the medical implications of nuclear weapons.
- Seek the cooperation of the medical and related professions in all countries for these aims.
- Develop a resource center for education on the dangers of nuclear weapons and nuclear war.
- Encourage studies of the psychological obstacles created by the unprecedented destructive power of nuclear weapons which prevent realistic appraisal of their dangers.
- Initiate discussion of development of an international law banning the use of nuclear weapons similar to the laws which outlaw the use of chemical and biological weapons.
- Encourage the formation in all countries of groups of physicians and committees within established medical societies to pursue the aims of education and information on the medical effects of nuclear weapons.
- Establish an international organization to coordinate the activities of the various national medical groups working for the prevention of nuclear war.

Excerpt from abstract of IPPNW Proceedings of First Congress held in Airlie, Virginia, on March 20-25, 1981.

19 Epilogue

PETER G. JOSEPH

Physicians for Social Responsibility began in the early 1960s, focusing on the dual issues of atmospheric nuclear fallout and the fallacy of civilian defense against nuclear attack. The organization now has over 40 chapters and 5,000 physician members in the United States, with new chapters forming around the world. We intend to continue our vigorous efforts to alert the public, the medical profession and our elected representatives about the health consequences of nuclear war.

Why do physicians and scientists choose to focus on the details of how the world might end? For one, physicians are constantly confronted with bad news, with difficult matters of individual life and death. The subject of nuclear war confronts each of us with intense feelings about the meaning of our lives, and of our deaths. We know the present situation is grave, the prognosis guarded; but we also know that in medicine there is always hope, for the annals of medicine are full of tales of critically ill patients who have outlived their pessimistic physicians. Hope is a prerequisite for medical practice. It is an absolute necessity for functioning in the nuclear age.

Physicians who deal with the life process from beginning to end, who appreciate its magnificence, its brilliant resiliency, its delicate requirements, understand that our species can leave a more distinguished message in the geological record than another mass extinction.

Scientists have confirmed something we have suspected all along: that as far as we know, we are all alone in this corner of the universe, hurtling through space on a small, fragile, isolated gem of a planet. As physicians and scientists, we feel a deep responsibility to protect from harm the biosphere which nourishes all life.

The health of *all* species, including ours, depends on its integrity.

And we recognize our obligations to others: Thousands of generations of humans have worked hard to continue our species and pass along an unspoiled world. We owe it to our ancestors to continue that work. Our children deserve to inherit an intact physical and genetic environment. And Earthlings, including ourselves, deserve better than to be incinerated in a nuclear war. May our children's children be able to thank us for choosing the path which leads toward life.

"At the present moment of history, there must be a general mobilization of all men and women of good will. Humanity is being called upon to take a major step forward, a step forward in civilization and wisdom. A lack of civilization, an ignorance of man's true values, brings the risk that humanity will be destroyed. We must become wiser."

—Pope John Paul II

Contributors

HERBERT L. ABRAMS, M.D., is Philip H. Cook Professor of Radiology at Harvard Medical School and Chairman of Radiology at the Brigham and Women's Hospital and the Sidney Farber Cancer Institute. The author acknowledges the assistance of William E. Von Kaenel and the support of the Henry J. Kaiser Family Foundation.

HELEN M. CALDICOTT, M.B., B.S., is President of Physicians for Social Responsibility. Caldicott is an Associate in Medicine at Children's Hospital Medical Center, and Instructor in Pediatrics at Harvard Medical School. She is the author of *Nuclear Madness, What You Can Do* (1979).

EVGENI I. CHAZOV, M.D., is the USSR Deputy Minister of Health, the Director-General of the National Cardiology Research Center and a member of the Presidium of the USSR Academy of Sciences.

JOHN D. CONSTABLE, M.D., is Associate Clinical Professor of Surgery at Harvard Medical School, Visiting Surgeon at the Massachusetts General Hospital and Chief Consultant in plastic surgery at the Shriners Burns Institute in Boston.

BERNARD T. FELD is Professor of Physics at the Massachusetts Institute of Technology and Editor-in-Chief of the *Bulletin of the Atomic Scientists.* He is the co-editor of *The Collected Works of Leo Szilard: Scientific Papers* (1972), and *New Directions in Disarmament* (1981), and the author of *A Voice Crying in the Wilderness: Essays on the Problems of Science and World Affairs* (1979).

STUART C. FINCH, M.D., is Chief of the Department of Medicine at the College of Medicine and Dentistry, Rutgers Medical School at Camden. From 1960 to 1962 Finch was Chief of Medicine at the Atomic Bomb Casualty Commission, and Special Consultant to the Director of the Commission from 1974 to 1975. From 1978 to 1979 he was Vice Chairman and Chief of Research at the Radiation Effects Research Foundation in Hiroshima.

ROGER FISHER is Williston Professor of Law at Harvard Law School, and director of the Harvard Negotiation Project. He was the originator and executive editor of *The Advocates* (1969–1970) and *Arabs and Israelis* (1974–1975), public television series produced by WGBH. Fisher is the author of *International Crises and the Role of Law: Points of Choice* (1978) and *Improving Compliance with International Law* (1981), and co-author of *Getting to Yes: Negotiating Agreement Without Giving In* (1981).

JOHN KENNETH GALBRAITH is the Paul M. Warburg Professor of Economics Emeritus at Harvard University. He served as the U.S. Ambassador to India from 1961 to 1963. He is a former editor of *Fortune* magazine and the author of many books—*The Affluent Society* (1958), *The New Industrial State* (1967), *The Age of Uncertainty* (1977). His most recent book is *A Life in Our Times* (1981).

H. JACK GEIGER, M.D., is Arthur C. Logan Professor of Community Medicine and Director of the Program in Health, Medicine and Society in the School of Biomedical Education of the City College of New York. In 1973 he received the Rosenhaus Foundation Award from the American Public Health Association, and an award from the Mississippi Association for Community Health for the Poor. In 1979 he received an award from Blue Cross and Blue Shield Associations for "developing a new concept of community health care as an effective instrument of community change."

HOWARD H. HIATT, M.D., is Dean and Professor of Medicine at the Harvard School of Public Health.

JOHN DAVID ISAACS is the Legislative Director of the Council for a Livable World in Washington, D.C.

PETER G. JOSEPH, M.D., is President of the San Francisco Bay Area Chapter of Physicians for Social Responsibility. He is on the teaching staff of the Department of Internal Medicine of the University of California, San Francisco Medical Center.

GEORGE B. KISTIAKOWSKY is Professor of Chemistry Emeritus at Harvard University. He was Special Assistant for Science and Technology under President Eisenhower from July 1959 to January 1961 and a member of the President's Science Advisory Committee from 1957 to 1964. Kistiakowsky is Chairman of the Council for a Livable World.

ROBERT JAY LIFTON, M.D., holds the Foundation's Fund for Research of Psychiatry professorship at Yale University. He is the author of *Death in Life: Survivors of Hiroshima* (1976), *The Broken Connection* (1979), and *Six Lives/Six Deaths: Portraits from Modern Japan* (1979). The drawings in THE FINAL EPIDEMIC are by Lifton and from *Birds* (1968) and *PsychoBirds* (1978).

BERNARD LOWN, M.D., is Professor Cardiology at Harvard School of Public Health, and a physician at the Brigham and Women's Hospital in Boston. Lown is the founder and first president of Physicians for Social Responsibility, and founder and first president of International Physicians for the Prevention of Nuclear War.

PATRICIA J. LINDOP, FRCP, is Professor of Radiation Biology at St. Bartholomew's Hospital at the Medical College of London. She is chairman of the University of London Board of Studies in Radiation Biology.

JOHN EDWARD MACK, M.D., is Professor of Psychiatry, Cambridge Hospital, Harvard Medical School. He received the 1977 Pulitzer Prize in Biography for *A Prince of Our Disorder: The Life of T.E. Lawrence* (1976). He is the co-author of *Vivienne: A Study of Adolescent Suicide* (1981). Mack is "grateful to Jack Ruina, consultant to the American Psychiatric Association's Task Force on the Psychosocial Impacts of Nuclear Advances for starting me and other members of the Task Force thinking seriously about the psychological and emotional context of the nuclear arms race."

J. CARSON MARK, a mathematician, was the Director of the Theoretical Physics Division of Los Alamos Scientific Laboratories from 1947 to 1973.

JOSEPH ROTBLAT is Emeritus Professor of Physics at the University of London. He is the former Secretary-General of Pugwash, the past President of the British Institute of Radiology and the British Hospital Physicists' Association. For many years he was Editor-in-Chief of *Physics in Medicine and Biology* and for 28 years the chief physicist to St. Bartholomew's Hospital in London.

HERBERT SCOVILLE, JR., is President of the Arms Control Association, and a Board Member of the Council for a Livable World. Scoville was Technical Director, Armed Forces, for the U.S. Department of Defense (1948–1955), Deputy Director for Research of the Central Intelligence Agency (1955–1963), and Assistant Director for Science and Technology of the U.S. Arms Control and Disarmament Agency (1963–1969). He is the author of *MX: Prescription for Disaster* (1981).

VICTOR W. SIDEL, M.D., is Professor and Chairperson of the Department of Social Medicine, Montefiore Hospital and Medical Center and the Albert Einstein College of Medicine in New York. Sidel is President-Elect of the Public Health Association of New York City. He is an editor of *The Fallen Sky: Medical Consequences of Nuclear War* (1963), and co-author of *A Healthy State: An International Perspective on the Crisis in U.S. Medical Care* (1978), and *The Health of China* (in press).

KOSTA TSIPIS is Associate Director of the MIT Program in Science and Technology for International Security in the Physics Department of the Massachusetts Institute of Technology.

RUTH ADAMS is the Editor of the *Bulletin of the Atomic Scientists.*

SUSAN CULLEN is the Managing Editor of the *Bulletin of the Atomic Scientists.*

Index